United States Women in Aviation 1930–1939

Claudia M. Oakes

SMITHSONIAN INSTITUTION PRESS

Washington and London

Library of Congress Cataloging in Publication Data
Oakes, Claudia M.
United States women in aviation, 1930–1939.
Bibliography: p.
1. Women in aeronautics—United States. I. Title. II.
Series.
TL521.017 1985 629.13'092'4 85-600019
ISBN 0-87474-709-0

Manufactured in the United States of America
97 96 95 94 93 5 4 3 2

⊗ The paper used in this publication meets the
minimum requirements of the American National
Standard for Permanence of Paper for Printed
Library Materials Z39.48-1984

For permission to reproduce illustrations appearing
in this book, please correspond directly with the
owners of the works, as listed in the individual
captions. The Smithsonian Institution Press does not
retain reproduction rights for these illustrations
individually, or maintain a file of addresses for
photo sources.

ᏇᎧ

CONTENTS

United States Women in Aviation
1930–1939

Claudia M. Oakes

Introduction

The 1930s marked a positive change for aviation in general and for women in aviation in particular. Amelia Earhart, the best-known female pilot in the history of aviation, defined for the decade what women were trying to prove by their flying; it was a dual message—flying is safe, and women make good pilots. The two ideas went hand in hand.

Women used their smaller physique and their lesser physical strength to try to dispel the notion still lingering from World War I that pilots were larger-than-life "he-men." Manila Davis, who gave up an acting career in the 1920s to pursue a career in aviation, became a saleswoman for the Curtiss Wright Flying Service in Boston; she was quoted in the newspapers, giving one of the reasons for her employment by Curtiss: "If I can fly and land a plane successfully, weighing as I do but 105 pounds, almost anyone ought to be able to."[1]

Manila Davis was not the only one of her sex to become a saleswoman in aviation. In 1930, Bessie Davis of Brooklyn, New York, sold aircraft instruments for Pioneer Instrument Co. Louise Thaden, Blanche Noyes, and even Amelia Earhart are some of the better known women who worked in aircraft sales.

Women were also trying hard to demonstrate that they could learn to fly as quickly as men. Dana Thompson, chief pilot for Summit Flying Service in San Francisco in 1930, had few doubts. Women were "easier to teach, and learn quicker than men," he remarked. One of the reasons he gave was that they took flying more seriously. "Women usually think about flying for a long time before they start taking instruction . . . ," he noted. "They leave the instruction to you. When you tell them their mistakes they pay more attention and consequently correct them quicker" (Wiggins, 1930:9).

The 1930s also saw the entrance of women into the highly competitive spheres of air racing and commercial air travel. At first, women competed against each other in their own races, but by the mid-1930s, they were flying against male competitors in such prestigious events as the transcontinental Bendix Trophy Race. In 1936, Louise Thaden and her copilot Blanche Noyes won the Bendix, and Laura Ingalls finished second. Jacqueline Cochran repeated a first-place for women in the 1938 Bendix.

Leading women pilots also took part in the development of commercial air travel. Here they did much to boost the fledgling air transport industry by writing articles and giving speeches on the safety, convenience, and even luxury of air travel. In 1930, Boeing Air Transport hired the industry's first stewardesses. A major breakthrough for women followed in 1934 when Helen Richey was hired as a pilot for Central Airlines; unfortunately, her employment lasted only a few months because of pressure from male airline pilots.

By 1930, there were approximately 200 licensed women pilots in the United States (Author unknown, 1930a:44). By late 1935, that number had

Claudia M. Oakes, Department of Aeronautics, National Air and Space Museum, Smithsonian Institution, Washington, D.C. 20560.

[1] "Girl Quits Stage to Take up Aviation," unidentified, undated newspaper clipping, Manila Davis Talley scrapbook, National Air and Space Museum archives.

2

FIGURE 1.—Manila Davis' small stature did not prevent her from becoming a successful saleswoman for Curtiss Wright in Boston. (SI photo 79-9382)

FIGURE 2.—Charlotte Frye, shown here with her beautiful Staggerwing, served for many years as a saleswoman for Beechcraft. (Courtesy of Beechcraft Corp.)

FIGURE 3.—Caricatures of women pilots. (Courtesy of Fay Gillis Wells)

grown to between 700 and 800 (Studer, 1935:20). *National Aeronautics* magazine in mid-1935 surveyed the professions of 142 licensed women pilots in the United States, with interesting results. Twenty-one worked professionally in aviation as executives, instructors, ferry pilots, saleswomen, flying school operators, aerial crop surveyors, and transport pilots; thirty-five, however, listed their profession as housekeeper, wife, or mother; thirty-three had independent incomes, and one of these

listed her profession as "private amusement"; others included teachers, journalists, nurses, physicians, artists, social workers, and college students, as well as one private investigator, among other diverse professions (Studer, 1935:20).

Why did women want to fly? Many of their reasons were the same as those given by men who were attracted to aviation. The explanation given by Margery Brown (1930:30), a very eloquent spokesman for women in the 1920s and 1930s, was one with which aviators of both sexes could identify.

Why do I want to fly? Because half-way between the earth and sky, one seems to be closer to God. There is a peace of mind and heart, a satisfaction which walls can not give. When I see an airplane flying I just ache all over to be up there. It isn't for a fad, or a thrill, or pride.

But Brown went further, summing up feelings about aviation that were probably unique to women (Brown, 1930:30).

Women are seeking freedom. Freedom in the skies! They are soaring above temperamental tendencies of their sex which have kept them earth-bound. Flying is a symbol of freedom from limitation.

Brown believed that flying would make women

FIGURE 4.—From an unattributed newspaper clipping of 1930 in the collection of Babe Weyant Ruth comes this account of advice on female flying attire.

A few years ago we would have scoffed at the idea of including travel-by-air costumes in our wardrobes, but with airplanes rivaling trains in rates and outdistancing them in speed more and more travelers are taking to the air for transportation to summer playgrounds, thus gaining a few more precious holidays.

Should you take off in a luxurious cabin plane, fitted up with all the comforts of a pullman or ocean steamer, you will suit the occasion, so to speak, in tweeds, cut in tailored style. To go with a tweed suit, a smart blouse could be one made of a fine jersey, similar in style to a tuck-in pull-on, but a bit more formalized through such details as wooden or leather buttons at a side closing on the neck and cuffs.

A dress of one of the flat sheer wool crepes is an alternative to the tweed suit, and for short "hops" from city to city leaves you properly clad for your activities at both ends of the trip. The hat for such an outfit, of course, would be one of the chic little draped berets that have captivated Paris. A long coat, either matching the dress in color and material, or of a harmonizing tone in a heavier crepe or basket weave would be a wise addition, since one of the air's greatest summer attractions is its coolness.

So far we have made no mention of costumes for those whose interest in flying is not confined to playing the role of passenger. Sketched at the left is an aviatrix outfit that more than covers this situation. Of wool crepe, this suit is a three-in-one combination of blouse, culotte trousers and panel skirt all in one piece. Zipper fastenings on everything that fastens, and inverted pleats in the skirt and culottes are other devices to add to the practicality and comfort of feminine flying togs.

Helmets come in a light-weight gabardine, and needless to say, may be had in all the gayest of colors.

more self-reliant and more confident in their ability to make decisions. Women who flew would, she felt, also use while on the ground such piloting skills as being resourceful, thinking rapidly, and acting without delay. Aviation would foster in women greater independence.

There are many people connected with the Air and Space Museum whom I wish to thank for their assistance in compiling this study. I am particularly grateful to Walter J. Boyne, Director; Donald S. Lopez, Deputy Director; and E.T. Wooldridge, Chairman of the Department of Aeronautics, for their continuing advice and support. Others in the museum who were of great assistance were Kathleen Brooks-Pazmany, Dorothy Cochrane, Tom Crouch, Von Hardesty, Dom Pisano, and Jay Spenser. Sybil Descheemaeker, Susan Owen, and Elaine Fields labored long on the word processor to get this into print.

I also wish to thank Fay Gillis Wells, who was flying in Russia in the 1930s, Jean Ross Howard of the Aerospace Industries Association, and Pat Thaden Webb, daughter of Louise Thaden, for their unfailing support, advice, and interest.

There are four frequently mentioned names of women with whom I have been privileged to work. These four have all died during the course of my research for this volume, but they had all been very generous with their time to reply to my requests for assistance. I will always treasure my brief acquaintance with these outstanding aviation pioneers: Louise Thaden, Jacqueline Cochran, Jeannette Piccard, and Blanche Noyes. I dedicate this book to them.

Engineers, Instructors, and Entrepreneurs

One day in 1930, the secretary to the president of an airplane manufacturing company in the midwest announced to her boss that a representative of the Pioneer Instrument Co. of Brooklyn, New York, had arrived to see him. The representative's card read, "B. Davis." The company president was most anxious to meet the young "man" with whom he had been corresponding. When B. Davis walked into the office, however, it became obvious that the "B" did not stand for Ben or Burt but for Bessie (Albert, 1930:88).

The appearance of a woman selling aircraft instruments was not that unusual, however, since women were quite active in this endeavor, as well as in demonstrating and selling airplanes. Davis was not the only one of her sex engaged in professions that in the 1930s might at first glance appear to be all-male fields. One of many women so employed was Betty Huyler Gillies of New York. Gillies had received her private pilot's license in 1929 and by June 1930, had exceeded the 200 solo hours required for a transport license.

Gillies was hired first by Curtiss Wright Flying Service on Long Island in their sales department,

selling airplanes and flight courses. That entailed contacting prospective buyers, demonstrating aircraft performance, and giving talks on aviation to various groups in the area. The head of the Speakers Bureau at Curtiss Wright was Helen Weber, and Gillies also was often called upon to fly Weber to other cities for speaking engagements (Gillies, 1979:1).

In addition to Curtiss Wright, Gillies worked in a similar capacity for Waco Sales of New York; Country Club Flying Service of Hicksville, New York; and Gillies Aviation Corp.; the latter was operated in partnership with Gillies' brother-in-law. Her various duties with these companies included picking up new airplanes at the factories, charter flying, and some instructing.

The first flying school in the United States to organize a special women's division was the Penn School of Aviation in Pittsburgh. Louise Thaden took charge of this division when it was formed in 1930. Instruction was given at Pittsburgh-Butler Airport where male students were trained on the same equipment and by the same personnel. The separate division was created because the operators

FIGURE 5.—Betty Huyler Gillies worked in sales and demonstration flying for several companies including Curtiss in New York. (Photo from Curtiss Wright Collection, National Air and Space Museum)

of the school believed that methods for training women to fly should in some way be different from those used to train men, though no one seemed really sure why or how.

The results reported by Thaden at the end of the division's first year of operation disclosed that the cost of training male and female pilots was the same, both in ground and flying courses. Thaden reported that although women were generally slower in learning how to land, they were better in air work and quicker in learning takeoffs (Thaden, 1931:22).

In an article by another female flight instructor, Helen Johnson, words of advice were given to female instructors who might want to break a student's feeling of overconfidence without breaking their own ladylike image by resorting to "cuss words" (Johnson, 1935:10).

> She might let the student[s] know in no uncertain terms that they are poor students, that they are miserable flyers, that there is no use to proceed with instruction because they will never solo. Or, she might even put on a parachute, give the student one, and take out for high altitude. A routine of aerobatics until the student becomes ill often helps—in other words, one method is to frighten him.

Willa Brown, who received her pilot's license in 1937 and later became the first black woman to receive a commercial pilot's rating, also ran her own flying school. Brown had become interested in aviation in the late 1920s through reading about the aviation career of Bessie Coleman, the world's first black licensed pilot.

In the 1930s, Chicago was a major center for blacks who were trying to get into aviation. Brown joined the Challenger Air Pilots' Association, a flying club led by Cornelius R. Coffey and John C. Robinson and in 1934 began flying at Harlem Airport in Oak Lawn, Illinois, a Chicago suburb, earning her pilots's license in 1937.

She later organized a flying school at Harlem Airport and in 1939, with Cornelius Coffey, formed the National Airmen's Association of America. The purpose of the Association was to get blacks into the U.S. armed forces as aviation cadets. Brown also went to Washington, D.C., to lobby for blacks to be allowed into the Civilian Pilot Training program (Johnson, 1974:26).

By the end of the 1930s, women were also running their own charter services. For example, by 1939, Inez Gibson was the owner and operator of Union Air Services at the Union Air Terminal in Burbank, California, and had earned both a private and commercial license. She had learned to fly in 1936 under the instruction of Joe Lewis, owner of the Lewis Air Service in Burbank. Although aviation initially had been for her just a pleasant pastime, her attitude changed when Lewis was killed in October 1938 and his business put up for sale. Gibson bought the operation and the four Waco aircraft that went with it and changed its name to Union Air Services. She specialized in charter flying, aiming for the businessmen and the wealthy travelers who had become used to the convenience and quickness of traveling by air but who wanted to reach destinations that were not on an established airline route. Passengers were insured on every flight, and the aircraft were covered for property damage and personal liability.

Gibson was a shrewd businesswoman. She realized that people still had doubts about flying with women pilots, so she hired men to do the flying. "The answer is to give them a man pilot," she once said, "and avoid that much sales resistance" (Calkins, 1939:49).

She did not stop flying when she became an air service operator, however. Indeed, she began acquiring more advanced flying skills, studying navigation and instrument flying with the noted racing and test pilot Max Constant, and receiving aerobatic instruction from Tex Rankin, holder of numerous trophies for his racing and aerobatic flying (Calkins, 1939:50).

Another rather unusual activity for a woman pilot in the late 1930s was helping to popularize air mail. Harriet Davidson had soloed on 16 August 1935 and had received her pilot's license in 1937. Because her family had been opposed to her learning to fly, however, the money for her instruction had come solely from her salary as a medical technologist.

In May 1938, she participated in the attempt of the Post Office Department to popularize the use of airmail in New Mexico, which entailed commissioning private pilots to fly to small towns all around the state. Davidson, the only woman in the program,, was assigned to fly the first airmail in and out of Socorro, New Mexico, in a Fairchild 24.

Socorro did not have a runway, so the local highway department scraped one out, down the middle of a dry creekbed. Since most of the children in Socorro, and many of the adults, had never seen an airplane on the ground, the schools were closed the day Davidson flew in, and the whole town turned out to see her and her airplane (Nye, 1982:2).

FIGURE 6.—Janet Waterford Bragg, a registered nurse, was active in Chicago's black aviation community in the 1930s and purchased the first aircraft for the Challenger Air Pilots' Association, a group of air enthusiasts organized to interest blacks in aviation. (S.I. photo 79-13664)

FIGURE 7.—Lola Jones (left) and Willa Brown in 1935 at Harlem Airport, which was located at Harlem Avenue and 87th Street in Chicago. (S.I. photo 79-13665)

FIGURE 8.—Harriet Davidson. (Courtesy of Harriet D. Nye)

Later, Davidson bought her own Fairchild 24 and became the first woman in the state of New Mexico to own and fly her own plane. She was also the first member of the Ninety-Nines in New Mexico (Nye, 1982:2). (The Ninety-Nines is an organization of women pilots that was established in 1929.)

The first U.S. government program conceived, planned, and directed by a woman with an all-woman staff was the Bureau of Air Commerce's National Air Marking Program. In 1933 Phoebe F. Omlie was appointed Special Assistant for Air Intelligence of the National Advisory Committee for Aeronautics (forerunner of the National Aeronautics and Space Administration). The following year, she convinced the chief of the Airport Marking and Mapping Section of the Bureau of Air Commerce to institute a program whereby each state would take part in a program to better identify its towns and cities from the air.

The state would be blocked off in sections of 20 square miles. Wherever possible, a marker with the name of the nearest town would be painted on the roof of the most prominent building at each 15-mile interval. Where towns were too far apart, it was suggested that ground markers of rocks or bricks painted white be used (Thaden, 1936:15).

At that time, few private pilots were flying over established airways or with benefit of radio. With the aid of the markers, even the most inexperienced pilots could determine where they were, should they become lost or confused.

The program was funded by a system of state grants from the Works Progress Administration. Not only was it the first appropriation of funds specifically set up to aid private pilots (Thaden, 1936:15), but it was also hoped that the program would provide jobs for the unemployed and would establish valuable permanent airway aids. By mid-1936, 30 states were actively involved in the program, with approvals given for 16,000 markers at a cost of about one million dollars (Thaden, 1936:14).

In 1935, Omlie chose five leading women pilots as field representatives for the program: Louise Thaden, Helen Richey, Blanche Noyes, Nancy Harkness, and Helen McCloskey. These women were by that time well known in aviation.

At the beginning of the 1930s, Louise Thaden was employed as a public relations director of Pittsburgh Aviation Industries and director of the women's division of the Penn School of Aeronautics, while continuing her record-setting flights that had begun in the late 1920s.

In August 1932, Thaden and Frances Marsalis set a women's endurance record of 196 hours, 5 minutes at Curtiss Field, Long Island, in a Curtiss Thrush. Their friend Viola Gentry, also a well-known pilot, was in charge of food and supplies, and Charles A. "Casey" Jones was their flight manager. They exceeded by 74 hours the previous women's endurance record, which had been set by Bobbi Trout and Edna May Cooper in 1931. Because of the publicity generated by their flight, they were urged to break their pattern over Long Island, fly to Cleveland to open that year's National Air Races, and then fly back to New York to land. They landed earlier, however, because "the boredom became too strong" (Thaden, 1938:129).

Although the press dubbed their aircraft the "Flying Boudoir" because they once requested hairpins and cold cream, Thaden and Marsalis did not have comfortable conditions. They had stripped their six-place aircraft of all but the pilot's and copilot's seats. Their only air mattress was punctured during their second day aloft, and Marsalis

FIGURE 9.—Air Marking pilots (left to right) Blanche Noyes, Helen Richey, Helen Mc-Closkey, Louise Thaden. (Courtesy of Louise Thaden, S.I. photo 79-7394)

suffered from severe back pains for three of the eight days of the flight.

Thaden followed her endurance record with two speed records, one in July 1934 for a light plane speed record of 109.58 miles per hour over a 100-kilometer course in a 90-horsepower Porterfield, and an east-to-west speed record in August 1936 in a Beechcraft Staggerwing. The latter was a prelude to what was to become Thaden's most famous aviation achievement. In 1936, with Blanche Noyes as her copilot, she became the first woman to win the Bendix Trophy Race. She later described some of the problems she had during that flight (Sutton, 1970:3).

Our navigational helps were not very good. We had a receiver but it went out and we had to just dead reckon. I had a knee pad and every time we had to deviate off our compass course I would write down the time and our heading and then the first time that we found a valley going back in that direction, I'd write that time and heading. In that way we stayed on course all the way to Los Angeles.

Thaden was also awarded the Harmon Trophy as the leading female pilot of 1936.

Besides her flying, Thaden played an active role in promoting the aviation industry through her speeches, her articles, and her autobiography, which was published in 1938. She also served as vice-president of the Ninety-Nines in 1931–1932, the first year that group had a full slate of officers (Thompson, 1970:6), and as National Secretary of the National Aeronautic Association in 1937–1938.

At the end of the 1930s, Thaden returned to the commercial side of aviation, serving as a factory representative and demonstration pilot for Beech Aircraft Corporation.

FIGURE 10.— Nancy Harkness, one of the women involved in the Air Marking Program, later became head of the Women's Auxiliary Ferrying Squadron of the U.S. Air Transport Command. (S.I. photo A45968)

Louise Thaden summed up her thoughts on flying in the foreword to her autobiography *High, Wide, and Frightened* (1938).

A pilot who says he has never been frightened in an airplane is, I'm afraid, lying.

Pilots have an irresistible habit of doing two things: enlarging situations which occur to them while flying, and pooh-poohing fright.

Doubtless I shall be criticized by them for attempting to set down sensations and thoughts we have all experienced sometime or other, as well as for shattering the impression that a pilot is a super-individual full of iron nerve, remarkable courage, of calm efficiency which nothing can disrupt, of absolute control over brain and body.

Actually we are human beings with usual inhibitions, phobias, and frailties common among men. We are unusual only because of our constant devotion to and fierce defense of aviation.

In looking back on aviation in the 1930s from a perspective of 40 years later, Thaden described it as the first time women began to be accepted on their own merits as pilots. It was a time of growth and exploration, when all "firsts" were *really* firsts, a time when camaraderie existed because words were not always necessary between fellow pilots, a time of instant friends and a spirit of cooperation, and most of all, a sense of something shared (Thaden, 1979b).

Helen Richey joined the Air Marking Program after being disappointed in her first choice for an aviation career. Too often overlooked in an era of more-recent "firsts" is the fact that in December 1934, Helen Richey became the first woman to pilot a commercial airliner on a regularly scheduled route.

Richey's ambition to become a pilot began in 1929, when she and a friend bought tickets for an airplane ride between McKeesport, Pennsylvania, and Cleveland. They made the flight in a Clifford Ball Airlines' Waco biplane, perched on a stack of mailbags (Kerfoot, 1978:17).

Bad weather prevented the women from returning for several days, and they were allowed to stay around the Cleveland airport, sometimes convincing pilots to take them for rides. Richey became convinced that she, too, wanted to fly.

On 3 April 1930, she enrolled in the Curtiss-Wright flying school at Bettis Field near McKeesport, soloed on 29 April after six hours of instruction, and on 28 June became the first licensed female pilot in the Pittsburgh area. Perhaps a portent of what she would later do, Richey was asked to serve as copilot on a Pittsburgh Airways' Ford trimotor on 4 July 1930, for an airport dedication at Moinessen, Pennsylvania (Kerfoot, 1978:18).

Recognizing that the air transportation industry was not yet ready to accept women as commercial pilots, Richey became a stunt pilot and flew her first show in the late summer of 1930 at a meet in Johnsonburg, Pennsylvania. Her performance there led to a job at Bettis Field, entertaining visitors with her aerobatics. On 30 December, the Bureau of Air Commerce granted her a limited commercial license, which allowed her to carry passengers within a 15-mile radius of her base (Kerfoot, 1978:18).

In 1931, her father made her a present of a four-place Bird monoplane, and Richey began making a name for herself in air racing. In December 1933, she and Frances Marsalis set a women's endurance record of 9 days, 22 hours over Miami.

After Marsalis was killed during the 1934 Women's Air Meet in Dayton, Richey decided to give up air racing and stunt flying to try for her real dream—becoming a commercial airline pilot. She applied for a job with Central Airlines, which at the time was competing with Pennsylvania Airlines for the Washington-Detroit airmail contract. Central won the bid, and in order to further compete with

FIGURE 11.—Louise Thaden and Frances Marsalis' Curtiss Thrush (bottom) is fueled over Long Island by a Curtiss Robin during Thaden and Marsalis' endurance flight in August 1932. (Courtesy of Louise Thaden, S.I. photo 79-7393)

FIGURE 12.—Louise Thaden (left) and Frances Marsalis (right) are surrounded by admiring crowds after their endurance record of 196 hours, 5 minutes. In the center is Viola Gentry, who handled food and supplies. (Courtesy of Louise Thaden, S.I. photo 79-7392)

Pennsylvania Airlines for mail contracts and passenger service, Central hired Richey as a copilot. Of the nine applicants, she had been the only female. Richey realized that her fame as a racing and stunt pilot and the novelty of a woman in the cockpit had won her the job.

On 31 December 1934, Richey made her first flight for Central from Washington, D.C., to Detroit, in a Ford trimotor. Richey continued as a pilot for Central until October 1935. During those months she became increasingly aware that she was being used more as a public relations person than as a pilot, that she was spending more time on the lecture circuit than in the cockpit. Because she was a woman, the Bureau of Air Commerce had told Central not to let Richey fly in bad weather, and the all-male pilots' union rejected her membership application (Kerfoot, 1978:18). Very discouraged, Richey resigned from her job. In ten months with Central, she had only been allowed to fly about a dozen round trips.

A storm of protests broke out from women pilots, one of the loudest being from Amelia Earhart. In response, Omlie asked Richey to join the Air Marking Program, where she served through 1937.

Richey did not give up other aspects of her aviation career, however. In 1936 at Langley Field, Virginia, she set a new Class C (under 1200 pounds) light-plane speed record, flying the 100-kilometer course in 55 minutes at an average speed of 72 miles per hour in her Aeronca C-2. Several months

FIGURE 13.—Louise Thaden (above) with the bright blue Beechcraft Staggerwing (below), which she piloted to victory in the 1936 Bendix Trophy Race. (Courtesy of Louise Thaden, S.I. photos 79-7387 and 79-7390)

FIGURE 14.—Frances Marsalis (left) and Helen Richey with their Curtiss Thrush *Outdoor Girl* in which they set a women's endurance record of 237 hours, 42 minutes over Miami, Florida, 20–30 December 1933. (Photo from Rudy Arnold Collection, National Air and Space Museum)

later, she set a new altitude record for that class airplane of 18,000 feet.

Amelia Earhart invited Richey to accompany her as copilot in the 1936 Bendix Trophy Race. They were delayed by bad weather and mechanical problems and finished fifth, but they were pleased that their friends Louise Thaden and Blanche Noyes won the race.

By the end of the 1930s, Richey had turned her considerable talent to yet another aspect of aviation: flight instruction. Through her racing, her aerobatics, her commercial flying, and her teaching, Richey ranked high among her peers in the cause to advance aviation as an industry and to the acceptance of women as pilots.

Blanche Noyes, another of the women picked for the National Air Marking Program, had learned to fly in 1929, becoming the first licensed woman pilot in the state of Ohio. Like many of the women of that era, Noyes became interested in aviation through her husband, Dewey Noyes, who taught Blanche to fly. She soon joined Dewey, a pilot for Ethyl Corporation of New York, in a career in aviation. In 1930, at Ormond Beach, Florida, the Noyes took John D. Rockefeller, Sr., then over 90, for his first and only airplane ride. During that same year

Noyes began making a name for herself in aviation with her aerobatics, once setting a record for the greatest number of consecutive turns in a spin.

In 1931, Blanche Noyes became a pilot for Standard Oil of Ohio, flying a Pitcairn autogiro with the Standard logo. Her racing career also began that year. She did well in closed-course events at the All-American Air Races in Miami, and later that year she competed in the Sweepstakes Derby from Los Angeles to Cleveland, one of the first races in which male and female pilots competed against each other. Noyes came in sixth in a field of 35.

From 1932 to 1935, Noyes continued as a demonstration pilot for various aviation-related corporations and began what would be a lifetime of lecture tours promoting aviation and women as pilots. Her husband taught her instrument flying, and the two began considering an around-the-world flight. That dream and others were shattered, however, in December 1935, when Dewey Noyes was killed in the crash of his Beechcraft Staggerwing. Friends, such as Amelia Earhart and Louise Thaden, tried to comfort her, but it was ultimately through aviation that Noyes was able to find solace.

Noyes joined the Air Marking Program and became one of its most ardent supporters. When federal funds for the program ran out, she flew all over the country to help gain financial support from local chambers of commerce and civic groups.

Noyes continued in the Air Marking Program throughout the rest of the 1930s and was eventually named chief of the Air Marking Staff, a position she was to hold for almost three decades, not retiring until 1972. She spent her life promoting safety in the air.

During this period several women combined their flying careers with careers in journalism. One such American woman pilot did most of her flying in Russia while working as a journalist. Fay Gillis had already established a name for herself in aviation as a member of the Caterpillar Club (those who had parachuted from a disabled aircraft to save their lives) and as a charter member of the Ninety-Nines. When her father, a mining engineer, was transferred to Moscow in 1930 and took his family along, Gillis became a special free-lance correspondent in the Soviet Union for the *New York Herald Tribune*, the Associated Press, and various aviation magazines. Through her connections as a reporter and as a pilot, in 1933 Gillis became the first American woman to fly a Soviet-made civil aircraft and later

FIGURE 15.—Blanche Noyes. (S.I. photo 79-3162)

FIGURE 16.—Blanche Noyes (right) with her friend Amelia Earhart.
(S.I. photo A2117)

FIGURE 17.—During a trip to South Carolina in 1937, Blanche
Noyes was photographed indulging in another sport. (Courtesy of
Dexter Martin)

CURTISS FLYING SERVICE
PHOTO DIVISION

FIGURE 18.—Fay Gillis worked for the New York office of the Curtiss Wright Flying Service before moving to Russia. (Photo from Curtiss Wright Collection, National Air and Space Museum)

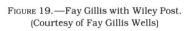

FIGURE 19.—Fay Gillis with Wiley Post. (Courtesy of Fay Gillis Wells)

.ТЕЛ

ПРИЕМ го ч. м. пор. № От пров. № Принял

45

ПЕРЕДАЧА го ч. м. пор. № В пров. № Передал

162 NEWYORK 128 48 18 5 52 =NLT FAY GILLIS ANANEVSKY PER DOM 5

№ сл. го ч м. APT 26 NSK =

Служебн. отметки

=LIKE YOU ARRIVE NOVOSIBIRSK BY JULY FIRST ARRANGE GAS PLANE IN TW
HOURS WHILLE I SLEEP THEN FLY WITH ME KHABAROVSK TO DIRECT
SERVICE THERE GET ME BEST MAPS NOVOSIBIRSK KHABAROVSK WILL
PAY YOUR EXPENSE REGARDS =WILLEY POST

СТАТ.

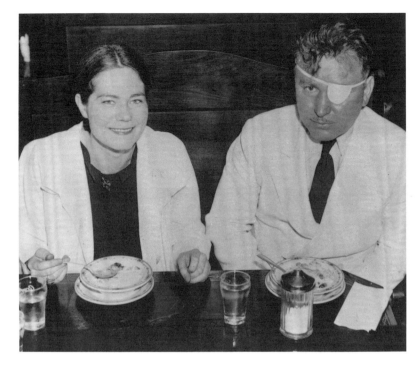

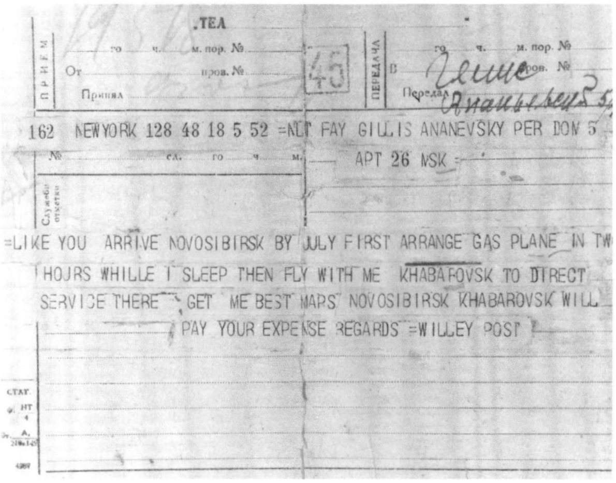

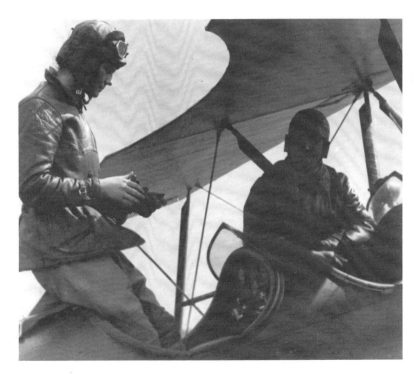

FIGURE 20.—Telegram sent by Wiley Post to Fay Gillis in 1933. (Courtesy of Fay Gillis Wells)

FIGURE 21.—Fay Gillis propares to take off in a Poliakarpov PO-2, 1933. (Courtesy of Fay Gillis Wells)

18

was the first foreigner to own a glider in the Soviet Union. In that same year, Wiley Post asked Gillis to handle the maintenance and refueling logistics in Russia during his solo round-the-world flight. Post considered her an ideal choice because of her aviation background and her knowledge of the Russian language. Gillis left Moscow on 2 July to meet Post at his first Russian stop, Novosibirsk. There she arranged for fuel, oil, and repairs for Post's plane, the *Winnie Mae*, and food and sleeping accommodations for Post himself. Wells had to have the grass on the airfield cut twice before Post showed up (Cowen, 1983:C2).

Wiley always kept the exact date of his flights secret, so he didn't have to apologize if they failed. I ended up waiting in Siberia for three weeks, flying out to Novosibirsk in a mail plane. They put me in the back and piled the mail around me.

Gillis was supposed to fly the next leg of the trip, 2500 miles to Khabarovsk, in a tiny improvised seat behind the gas tank in the *Winnie Mae*'s tail. However, it was decided that this would jeopardize Post's claim to a solo flight, so Gillis remained in Novosibirsk.

Post contacted Gillis again in the summer of 1935 asking her to be his copilot on his proposed flight from Los Angeles to Moscow. Post's planned route would take them first to Seattle, then to Alaska, across the Bering Strait to Siberia, and on to Moscow. Gillis, however, had married the journalist Linton Wells in the spring of 1935. Near the time of her planned flight with Post, her husband was hired by the *New York Herald Tribune* to cover the Italo-Ethiopian War, whereupon he proposed a delayed honeymoon. "I knew Wiley could replace me on the trip, but I didn't want anyone to replace me on my honeymoon," she later said (Cowen, 1983:C2). When Gillis opted for Ethiopia, Post took along Will Rogers instead, and the two were killed when their Lockheed Orion crashed near Point Barrow, Alaska.

Besides continuing her aviation career throughout the 1930s, Gillis with her husband covered the Italo-Ethiopian War in 1935–1936 and the Syrian riots in 1936 for the *Herald Tribune* and served as the *Tribune*'s special Hollywood correspondent in late 1936 and 1937. In 1939, she became a founding member of the Overseas Press Club.

One of the most famous women in aviation, who is also known as much by her writing as by her flying, is Anne Morrow Lindbergh. After her marriage in 1929 to America's most famous aviator, Anne Lindbergh lost little time in adopting her husband's pursuit of aviation. In early 1930, she became the first woman in the United States to earn a glider pilot's license; her teacher was the soaring pioneer Hawley Bowlus. Later that year, on 20 April, she served as copilot and navigator when Charles Lindbergh set a new transcontinental speed record of 14 hours, 23 minutes, 27 seconds.

In 1931, Anne Lindbergh received her private pilot's license. The young couple must have rejoiced in the solitude of flying together, because it was almost the only way in which they could be alone. Everywhere they went, they were surrounded by admiring crowds.

Perhaps the most famous flight the Lindberghs made together was the one that Anne Lindbergh described so eloquently in her book *North to the Orient* (1935), a flight that showed the feasibility of using the Great Circle Route to reach the Far East. They left Maine in their Lockheed Sirius in late July 1931 and followed a route that took them to Canada, Alaska, Siberia, the Kurile Islands, and then to Japan and China. On the flight, Anne served as copilot, navigator, and radio operator.

The Lindberghs' next major aviation venture occurred in 1933, while Charles Lindbergh was serving as technical advisor to Pan American Airways. Pan Am was participating with four other international airlines in a cooperative study of possible Atlantic commercial air routes. The Lindberghs' territory would be from Newfoundland to Europe via Greenland. Their aircraft once again was the Lockheed Sirius.

For her part in the study, Anne Lindbergh received the United States Flag Association Cross of Honor, and in 1934 became the first woman to receive the National Geographic Society's Hubbard Medal.

After completing their fact-finding transatlantic mission for Pan Am, the Lindberghs toured the major cities of Europe, went as far east as Moscow, flew down the west coast of Africa and across the South Atlantic and returned to the United States five and a half months after they had left.

During the 1930s, Anne Lindbergh also began her career as an author. In her first book, *North to the Orient* (1935:137–138), she described her fascination with aviation.

I was conscious again of the fundamental magic of flying, a miracle that has nothing to do with any of its practical purposes—purposes of speed, accessibility, and convenience—and will not change as they change . . . .
For not only is life put into new patterns from the air, but it is

FIGURE 22.—Anne Lindbergh in a Bowlus sailplane in early 1930; Charles Lindbergh is standing at near left. (S.I. photo 72-8713)

FIGURE 23.—Soaring pioneer Hawley Bowlus (left) taught Anne Lindbergh to fly a sailplane. (S.I. photo 72-8721)

FIGURE 24.—Charles and Anne Lindbergh with their Lockheed Sirius in 1931. (Photo from National Air and Space Museum)

FIGURE 25.—The Lindberghs' Lockheed Sirius in the harbor at Gothaab, Greenland, 1933.
(S.I. photo A45256J)

somehow arrested, frozen into form . . . . A glaze is put over life. There is no flaw, no crack in the surface; a still reservoir, no ripple on its face.

An unconventional aerial occupation for women, aerial police work, was attempted by the Los Angeles Chief of Police. In May 1937, he added five women to his squad of police pilots who could be summoned for duty whenever a situation arose in which their skills might be needed. The five were Mary Charles, Karena Shields, Bobbi Trout, Bettymay Furman, and Pretto Bell (Author unknown, 1937). Mary Charles had been the one who had promoted the idea with the Chief of Police, but, as Bobbi Trout wrote 46 years later, none of the five really were called upon to do anything with their aerial police work, and the idea finally died. "Those were the days," Trout said, "when flying did not pay much and many were flying for FREE just to keep up their licenses" (Trout, 1983:1).

The profession in aviation that is considered the most conventional for women—that of stewardess, now flight attendant—was also born in 1930.

Flying was not perceived as the most comfortable form of transportation in its infancy, and many people were apprehensive about traveling by airplane. Routes were often long, with frequent stops. Copilots did double duty. In the cockpit, they helped the pilot fly the plane; in the passenger cabin, they served meals bought from local restaurants before takeoff, answered questions, calmed nerves, and helped passengers who felt ill.

Early in 1930, Ellen Church, a registered nurse, visited the Boeing Air Transport office in San Francisco to discuss with the district manager her desire to become a pilot. Although she did not obtain a pilot position, Church did convince Boeing that hiring women as flight attendants would not only free copilots for their regular duties, but also would be a good way to show people that flying was safe. Thus, on 15 May 1930, the first group of eight stewardesses began service from Chicago in a Boeing 80A transport, westbound on the exhausting 20-hour, 13-stop route to San Francisco.

The first eight stewardesses were Church, Margaret Arnott, Inez Keller, Cornelia Peterman, Harriet Fry, Jessie Carter, Ellis Crawford, and Alva Johnson. The original requirements under which they were hired were that they be single, registered nurses, not more than 25 years old, weigh not more than 115 pounds, and not be more than 5 feet 4 inches tall. These first stewardesses were paid $125 per month for 100 hours of flying.

In addition to serving passengers, the stewardesses assisted in refueling the aircraft, transferring

FIGURE 26.—The first eight stewardesses, who were hired by Boeing Air Transportation 15 May 1930: (first row, left to right) Margaret Arnott, Inez Keller, Cornelia Peterman, Harriet Fry, Jessie Carter, Ellis Crawford; (second row) Ellen Church and Alva Johnson. (S.I. photo 78-14836)

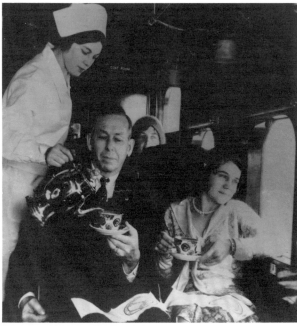

FIGURE 27.—In the 1930s, stewardesses' uniforms resembled nurses' outfits, and passengers were served coffee in china cups from silver pots. (Courtesy of United Airlines)

FIGURE 28.—Flight attendants looked for ways to keep their passengers entertained on the longer legs of their trips. In 1932, one method was the monitoring of the Hoover-Roosevelt presidential election. (Courtesy of United Airlines)

baggage, mopping the cabin floors, and checking bolts to ensure that all seats were securely fastened to the floor. Other inflight duties for stewardesses in 1930 were listed as follows (Wittman, 1975).

> Keep the cabin immaculate. Before flight, sweep the floor, dust off the seats and windowsills.
> Keep the clock wound up.
> Correct the time as the aircraft passes through time zones.
> Keep an eye on passengers when they go to the toilet room to be sure they go through the toilet room door and not through the emergency exit door.
> Carry a railroad timetable just in case the plane is grounded somewhere.

By the mid-1930s, there were between 200 and 300 stewardesses in airline service in the United States, and they had become an integral part of the air transportation industry. This profession was the only non-pilot one in aviation in which large numbers of women were employed.

However, there was some objection to using the word "stewardesses" to describe these first flight attendants. The editor of *The American Journal of Nursing* had this to say in a letter to Robert Johnson of Boeing Air Transport, who had hired the original eight stewardesses (Roberts, 1931:1).

> I know that other nurses have registered a protest against calling nurses stewardesses. Frankly, we are somewhat disturbed over this title. It is a less dignified term than nurse and would, I should suppose, lessen the chance of securing really good nurses for such service. Nevertheless, we are deeply interested in the development. We shall be grateful if you will inform us, promptly, of any expansion or other changes in the service.

At the start of the 1930s, *Western Flying* magazine conducted a survey of seven aircraft manufacturers to determine in what capacities they were employing women. Besides the usual clerical jobs, there were only two areas in which women were working: five of the plants were using women in the wing and fuselage covering departments; in one, women were used in light cadmium plating work. Of the 852 employees at the Boeing Company in Seattle, only 17 were women (Author unknown, 1930c:33).

The average weekly wage for the women who worked in these aircraft plants was less than that received by men for the same work. Men in the fabric covering departments earned 60–65 cents per hour for an average of $30 per week; women doing the same work received 40–50 cents per hour for an average of $21.50 per week. In figures for overall salaries in the aviation industry, there was an even greater disparity. The hourly average earnings for men were 66.9 cents, and for women 36.7 cents (Author unknown, 1930c:33).

Women still worked principally in the fabric shop areas of aircraft manufacturing plants in 1936. That year Clementina Griffin, principal of Narbonne High School in Los Angeles, decided to broaden her technical knowledge by taking an aviation mechanics course at Curtiss-Wright Technical Institute. Besides learning such "hands-on" skills as rib construction, welding, riveting, and cable splicing, Griffin's class took field trips to aircraft plants in the Los Angeles area. In an article describing the course, Griffin (1938:17) expressed her dismay at the dearth of women working in the mechanical side of aviation.

> In our trips through the factories I observed women only in one instance. A small group were working in a fabric department. There is no particular reason for or against women entering this industry. With the increase in the number who pilot planes there will naturally be more interested in the work. I believe that knowledge of the mechanics of the plane should precede the issuance of a license to a pilot.

By the end of the 1930s, there were several prominent female aeronautical engineers in the United States. Among them was Elsa Gardner who, in 1938, was editor of the prestigious *Technical Digest* published by the U.S. Army Air Corps at Wright Field in Dayton, Ohio.

Gardner became interested in pursuing a career as an aeronautical engineer while in college during World War I. She took a job in the gauge inspection division of the U.S. branch of the British Ministry of Munitions of War, and was later given an entry level position in the engineering section. She subsequently went to work for the U.S. Army Air Corps inspecting gauges used in aircraft and engine manufacture, and then was transferred to the Navy and placed in charge of all new master gauges used in the manufacture of torpedoes.

By the late 1920s, Gardner was set on becoming an aeronautical engineer. During the day she worked at the Eclipse Aviation Corp. in New York and at night attended the Pratt Institute and New York University. In 1932, Gardner won a partial scholarship to the Massachusetts Institute of Technology in the aeronautical engineering department. She further financed her schooling by writing for *Aero Digest*, a popular aviation magazine, of which she later became an editor.

Gardner became one of the very few women in that period who were elected to membership in the American Society of Mechanical Engineers, and was made a member of that organization's aeronautics committee, among whose other members

FIGURE 29.—Curtiss ground school students at New York University in New York City in the early 1930s. (Photos from Curtiss Wright Collection, National Air and Space Museum)

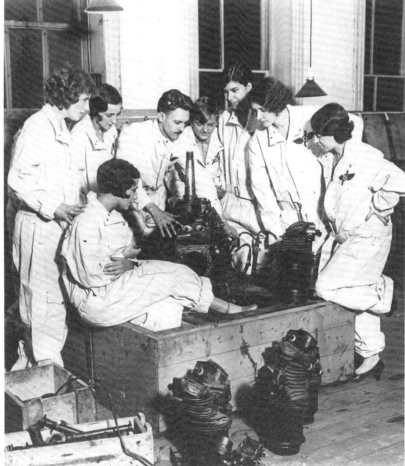

was Elmer Sperry, a pioneer in aircraft instrumentation systems.

Other U.S. female aeronautical engineers in the late 1930s included Mabel Rockwell, Dr. Frances Hurd Clark, Elsie Gregory McGill, and Isabel Ebel. Ebel, who in 1932 had laid out the course of Amelia Earhart's record transcontinental flight, succeeded in convincing the New York Board of Education that aeronautics should be taught to both girls and boys in New York high schools (Author unknown, 1938:13).

Toward the end of the 1930s, as war clouds were forming over Europe, the idea occurred to many in the United States that American manpower might be needed for national defense, so American womanpower would be in more demand to keep the country and the aircraft industry running. Women were encouraged to educate themselves for engineering and executive positions and to fill vacancies on aircraft assembly lines. Thus was born "Rosie the Riveter."

This drive was also intended to convince women pilots that their greatest contribution would be on the home front, not over the front lines as combat pilots. "Women should be builders, not destroyers," was one author's slogan (Schimmoler, 1939:34), and women pilots in the United States were encouraged to ignore "glamorous stories about women in other nations and their preparation for war," and to "cherish all the more the responsibility that has always been theirs, namely, that of the invisible guiding spirit" (Schimmoler, 1939:34). But that "invisible guiding spirit" was actually very visible in the 1930s, as women who were not even pilots used their skills to further the causes of the air transportation industry. Women had become an integral part of that industry.

Air Transportation

In the early 1930s, the fledgling air transportation industry began to see the advantage of women as airline passengers and began to make strong bids for their patronage. Among those women used to interest other women in air transportation was Amelia Earhart, who in 1928, as a passenger, had achieved her first "first"—the first woman to cross the Atlantic by air. During that flight in the Fokker F.VII *Friendship,* she said she had felt "like a sack of potatoes" because she had done none of the flying. But no matter how she felt herself, that flight marked the beginning of the American public's fascination with her, a fascination that still lasts, almost five decades after her disappearance.

Earhart's fame began immediately after the *Friendship* flight, and so did her serious interest in aviation. For Earhart, flying was never a casual sport or pastime. She was constantly training, learning, writing—all with the idea of proving that flying is safe and that women make good pilots. Her speed, altitude, and distance records, though numerous, were secondary to her goal of recognition for aviation as a viable means of transportation and for women as professional pilots.

On 21–22 May 1932, Earhart realized her goal of becoming the first woman to fly solo across the Atlantic, from Newfoundland to Ireland, in a Lockheed Vega. This flight made her the first person to fly across the Atlantic twice. Because her flight was five years to the day after Lindbergh's and because there was a perceived physical resemblance between them, Earhart was dubbed "Lady Lindy," a nickname she did not care for. Her genuine modesty and shyness, however, made her all the more popular with the press and the public, and she and her husband, George Palmer Putnam, were constantly in the limelight. She used this publicity to keep up public interest in aviation, particularly during the Depression. Among her best-known record flights were the first solo transcontinental crossing by a woman, from Los Angeles to Newark in her transatlantic Vega on 24–25 August 1932, the first solo flight by anyone from Hawaii to the U.S. mainland, 11 January 1935, in another Vega, and the first nonstop flight from Mexico City to Newark in May 1935 in her transpacific Vega. In addition to her flying, Earhart served as president of the Ninety-Nines, vice president of the National Aeronautic Association, assistant general traffic manager of Transcontinental and Western Air, and a member of the Guggenheim Committee for Aeronautical Education in Primary and Secondary Schools.

During 1936 and 1937, Earhart formed a close alliance with Purdue University and became a vis-

FIGURE 30.—Amelia Earhart with Irish farm people after her landing in Londonderry, 21 May 1932. (Photo from Adolfo Villasenor Collection, National Air and Space Museum, S.I. photo 80-11042)

iting professor there. In July 1936, she was given the use of a $70,000 Lockheed Electra, a virtual flying laboratory. Earhart used the aircraft, which had a ceiling of 27,500 feet, to test and observe human reaction to flights at high altitudes.

In the meantime Earhart was also preparing herself and the Electra for what she said would be her last long-distance flight (Backus, 1982:202), a flight around the world. Not only would she be the first woman to make such a flight, she would also be the first person to do so as closely as possible to the Equator, where the distance is the greatest.

The trip was planned with meticulous care and was originally to have been traveled from east to west. A mishap at Honolulu on the first attempt—a ground loop that severely damaged the Electra—caused a postponement.

While the plane was being repaired, Amelia and her navigator, Fred Noonan, restudied global weather conditions, and concluded that a June departure would probably favor a west-to-east journey of 29,000 miles.

The silver Electra left Miami at 5:56 A.M. on 1 June 1937, and there followed a succession of well-executed arrivals in cities across Africa and Asia and finally, in Lae, New Guinea. On 1 July—one month, 22,000 miles, and some 146 flying hours later—Amelia Earhart and Fred Noonan were poised for the most hazardous part of their journey, a never-before-attempted 2,566-mile hop to the tiny mid-Pacific island of Howland.

The airplane was performing well and both fliers were fit, but there were some minor problems. Headwinds delayed takeoff, and Noonan was unable to get the perfect radio reception that would have permitted him to synchronize his chronome-

FIGURE 31.—Crowds surround Earhart's Vega in Oakland after her nonstop solo flight from Honolulu in 1935. (S.I. photo 73-4032)

ters, which were so essential for precise navigation.

On 2 July, they departed Lae. There were radio messages, at first ordinary, but later concerned, indicating that the Lockheed had missed its minuscule island target, was low on fuel, and was unable to receive transmissions known to be coming from the U.S. Coast Guard cutter *Itasca.* Then there was silence.

A massive sea-air search was launched, and 265,000 square miles were covered by ships and aircraft of the U.S. Navy. On 18 July, the effort was called off, and Amelia Earhart and Fred Noonan were officially presumed lost at sea.

Earhart's disappearance immediately set off controversy and speculation as to the true mission of the flight and the fate of herself and Noonan. Countless books and articles have been written on the

subject, and the truth may never be known. What is known is that Earhart succeeded well in her dual goals—the air transportation industry and women pilots were given a boost forward by Earhart's support through her speeches, her writing, and her record flights.

After Earhart's disappearance, her friend and fellow pilot Louise Thaden (1938:258-259) wrote of her:

Like the rest of us, Amelia had ambitions. Unlike most of us she had a definite notion of each progressive step toward the set goal. Never swerving from sight of the beacon ahead, she climbed the stairs, step by step. Discouragement, frustration, hundreds of smaller obstacles, but probably most of all loneliness, could not deter her ascending the high pinnacle of predetermined achievements.

It may seem incongruous, yet A.E.'s personal ambitions were secondary to an insatiable desire to get women into the air; and once in the air to have the recognition she felt they deserved

FIGURE 32.—Earhart with her Lockheed Electra in which she disappeared on 2 July 1937.
(S.I. photo A45874)

FIGURE 33.—Amelia Earhart with her husband George Palmer Putnam (left) and Eleanor Roosevelt. (S.I. photo 82-8681)

FIGURE 34.—Paul Mantz, Amelia Earhart, and Fred Noonan in Hawaii in early 1937 before Earhart's first attempt at an around-the-world flight. (S.I. photo A38969)

accorded them. I have known many many women pilots she has helped either through financial assistance or moral encouragement. Further, she has talked more people into the air, most of them as passengers, others as pilots, than any other individual in aviation today.

During her aviation career, Earhart wrote articles for such magazines as *The Sportsman Pilot, The Aeronautic Review,* and *Cosmopolitan,* urging women to encourage their husbands to fly instead of driving or taking a train on business trips, and suggesting that flying was the best method of travel for a family vacation. In one of her many such articles, Earhart (1930b:15) recounted an incident which showed that air travel did not then immediately come to mind when one thought of going somewhere.

While in California recently I accepted an invitation to make a twenty-minute after-dinner talk in a Middle Western City. The distance was about thirteen hundred miles, and, of course, I flew. I arrived about four in the afternoon, two hours or more before dinner. The first words which greeted me from those awaiting me on the field were:

"Too bad you're so late. You missed a luncheon given in your honor this noon."

"I'm very sorry," I replied, "I'm *sure* I didn't receive notice of it. The message must have gone astray."

The chairman of the arrangements committee was found and she was honest.

"Well," she said, "I looked up all the trains from California and they all arrived in the morning, so I took it for granted you'd be here in time. I really didn't send you an invitation."

To her air travel had been until that moment so remote from everyday experience that she did not consider it. The conception of flying as hazardous war maneouvers [sic] or circus stunts at county fairs overshadowed that of transportation.

Certain amenities of passenger aircraft and more tasteful decoration of airport terminals may perhaps be attributed to the growing number of women airline passengers. Although these amenities were originally introduced to attract women to using air transportation, they equally attracted male passengers. On this subject, Earhart wrote, "Men appreciate the things which women will not travel without, and as the amenities are introduced into flying, more and more men are using air transportation" (Earhart, 1930a:32).

By the mid-1930s, women were making long duration and around-the-world flights as passengers and selling their stories to magazines and newspapers. The most famous of these was Dorothy Kilgallen, who was a reporter for the *New York Journal* when she made her world trip in October 1936.

Kilgallen was competing with two male newspaper reporters to see who could make an around-the-world trip in the least time. She began her journey

FIGURE 35.—Clara Adams, famous "first flighter," arrives at Newark Airport after her flight around the world in 1939. (S.I. photo 84-1761)

aboard the airship *Hindenburg* at Lakehurst, New Jersey, on 1 October, with less than 24 hours notice that her paper wanted her to enter the race. Because of this, she had no reservations, and page one of the 1 October 1936, edition of the *Journal* carried a plea for any Americans flying in Europe or the Orient to give up their seats for Kilgallen if so asked.

Kilgallen's daily reports to the *Journal* were given page one coverage for the 24 days of her journey. "DOROTHY NEARS MANILA," proclaimed the headline of the *Journal*'s 13 October edition. Munich, Rome, Athens, Alexandria, Beirut, Bagdad, Dehli, Calcutta, Rangoon, Bangkok, Hanoi, Hong

Kong, Manila—magical names to *Journal* readers, and yet photos of Kilgallen in such places proved they were only a plane flight away. After being grounded for five days by a typhoon, she boarded the *China Clipper* for San Francisco. On 24 October she landed at Newark, thereby winning the contest.

Clara Adams of New York City beat Kilgallen's passenger record in July 1939, flying 25,000 miles around the world, all on regularly scheduled airline flights, in 16 days, 19 hours, 4 minutes. Adams had already gained fame as a "first flighter," being a passenger on the first flights of the *Graf Zeppelin* and the *Hindenburg* to the United States, the first round trip flight across the Pacific, the first commercial flight from New York to Bermuda, the first clipper flight between San Francisco and New Zealand, and the first flight of the giant Dornier Do-X between New York and Rio de Janeiro.[2] She began her around-the-world flight on the first flight of the *Dixie Clipper* across the Atlantic. The around-the-world air fare cost Adams $1935 (Author unknown, 1939). (A first-class around-the-world trip by air in 1984 would cost about $4400, according to a Pan Am spokesperson.)

When Adams landed at Newark on a United Air Lines flight from San Francisco on 15 July, a *New York Times* reporter could not resist a fashion commentary, noting that she was "clad in a tan-plaid

---

[2] "Queen of 'First Flights'," unattributed, undated clipping, Clara Adams biographical file, National Air and Space Museum archives.

tailored suit, made of Chinese silk, purchased in Hong Kong, and wearing a tan Panama straw hat, purchased in Rangoon" (Author unknown, 1939).

By 1938, there was even a National Air Travel Week, and to mark the occasion, Alice Rogers Hager, a magazine feature writer, spent 1–9 October 1938 flying on regularly scheduled airline flights around the United States, Canada, and the Caribbean, covering 18,708 miles in 130 hours flying time (Hager, 1939:27).

Hager's trip began from Newark at 12:25 A.M. on 1 October on TransWorld Airlines. Her week-long itinerary included two round trips to the West Coast and three "zig-zags" from the north and south borders: Los Angeles to Seattle, Fort Worth to Detroit, and Montreal to Havana. She flew on 13 different airlines and on equipment ranging from the Douglas DC-2 to the then-modern Lockheed Electra and Lockheed 14. Her ticket, by the end of the flight, was almost 7 feet long (Hager, 1939:27).

Although Kilgallen, Adams, and Hager did not become pilots, they nevertheless had an infectious love of flying. They enthusiastically told of the wonders of the world and the beauty of the countryside as seen from the air, a new experience for most in the 1930s and one that many people felt they would never have. Clara Adams' description of her world flight as "beautiful beyond description and sublime beyond the most vivid imagination of the human mind" (Author unknown, 1939) must surely have inspired many earth-bound travelers to follow her lead and take to the skies.

Women Fly the Races

Shortly after 5:00 P.M. on 4 September 1936, a bright blue Beechcraft Staggerwing C-17R landed at Mines Field in Los Angeles near a pylon with the name "BENDIX" on its side. The pilot was obviously trying to avoid the crowded grandstand, as the plane turned to taxi far down the field. Suddenly, the plane's crew noticed a large group of men running toward their aircraft.

"I wonder what we've done wrong now!" Louise Thaden later recalled saying to her copilot Blanche Noyes (Thaden, 1938:181). Indeed she might have asked that question, for their problems had begun 14 hours and 55 minutes earlier at Floyd Bennett Field, New York, when the starter lost his flag and

had to drop his handkerchief as a signal for them to take off. Plagued with radio, wind, and weather problems the whole way across the continent, they finally reached Los Angeles, only to cross the finish line from the wrong direction.

Thinking they were the very last in the race to land, the women wanted to appear as inconspicuous as possible, and they were dismayed by the crowd surrounding them. Finally, the women realized that the men were race officials coming to congratulate them, for they had just won the prestigeous Bendix Trophy Race. In the crowd were Clifford Henderson, organizer and director of the National Air Races, and industrialist Vincent Bendix, the sponsor of the

FIGURE 36.—Louise Thaden (left) and Blanche Noyes (right) receive congratulations from Vincent Bendix after their win in the 1936 Bendix Trophy Race. (Courtesy of Louise Thaden, S.I. photo 79-7389)

race, who escorted the women to the microphones.

For the first time the race for the coveted Bendix Trophy had been won by women. Thinking that women might not be able to compete too successfully in the grueling transcontinental contest, a special prize of $2500, which sounded to the women suspiciously like a consolation prize, had been established for the first woman each year to complete the race. So Louise Thaden and Blanche Noyes received $7000—the regular Bendix purse of $4500 plus the $2500 special award. To the further amazement of the crowd, Laura Ingalls in her Lockheed Orion crossed the finish line 45 minutes later to win second place. Amelia Earhart and Helen Richey finished fifth.

Although the first Bendix Trophy Race occurred in 1931, it was not until 1935 that women actually competed. In a letter written many years later, Louise Thaden credited Clifford Henderson with that decision. She wrote, "Cliff Henderson . . . initiated and managed the National Air Races. Cliff

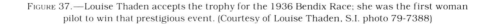

FIGURE 37.—Louise Thaden accepts the trophy for the 1936 Bendix Race; she was the first woman pilot to win that prestigious event. (Courtesy of Louise Thaden, S.I. photo 79-7388)

firmly believed women pilots were just as good as men pilots and it was he who made it possible for women to compete in this exciting annual event. Most importantly for women to compete with men in getting the fabulous Bendix Transcontinental Speed Race opened to women . . ." (Thaden, 1979a:1). If Clifford Henderson, "Mr. Air Races," said women could be allowed to compete, then that was all that was necessary. Up to that time air race officials simply did not believe that women pilots could—or should—compete against male pilots, so women were encouraged to hold their own competitions, such as the Women's International Free-For-All. Women were allowed on occasion to compete with men in some of the events at the National Air Races, but if a woman had an accident in such a competition, it was used as one more excuse to exclude female participation. Such was the reason given for keeping women out of the 1934 Bendix Race after Florence Klingensmith's fatal crash in the 1933 Frank Phillips Trophy Race in Chicago.

Frank Phillips, a wealthy aviation enthusiast from Oklahoma, had established a $10,000 prize and a trophy for the best time around a 100-mile triangular closed course. Klingensmith's aircraft was a Gee Bee Y. The Gee Bee racers, built by the Granville Brothers of Springfield, Massachusetts, were the quintessential racing aircraft of the 1930s. Their sometimes deadly speed was due to their short, bulky fuselages, clipped wings, and high-horsepower engines. The prestigeous Thompson Trophy Race had been won by Gee Bee racers in 1931 and 1932; Lowell Bayles flew to victory in the 1931 event in the Gee Bee Z, and the next year's Thompson was won by Jimmy Doolittle in the Gee Bee R-1.

Klingensmith was keeping up well with her male competitors, averaging over 200 miles per hour for 75 of the course's 100 miles, when spectators noted to their horror that the fabric on the plane's right wing was tearing loose. Klingensmith attempted to maintain control, but some of the fabric tangled in the tail surface. She left the pattern to avoid a collision with the other racers and to attempt a landing, but the plane crashed a mile and a half from the airfield. Her body was found tangled in her open parachute (Mandrake, 1957:64–65).

Although the crash was due to structural failure of the aircraft, Klingensmith's death was held up as an example of what could happen when women pilots were allowed to compete with men, and thus

FIGURE 38.—Florence Klingensmith. (Photo from Marjorie Stinson Collection, Library of Congress, S.I. photo 76-10579-34)

women were once again barred from the Bendix Race in 1934. The decision, however, did not pass without protest. Amelia Earhart refused to fly actress Mary Pickford to Cleveland to open that year's races (Dwiggins, 1965:79). Without any women in the closed-course events, either, the races that year lost a bit of spark.

Women held their own event in 1934, the First Women's National Air Meet, 4–5 August, at Dayton Municipal Airport, Vandalia, Ohio. Twenty women entered and participated in such events as 20- and 50-mile free-for-all races, precision landing contests, and bomb dropping. A parachute jumping contest was also scheduled but was cancelled because there was only one entry, Lucile H. Parker of South Charleston, West Virginia. The Contest Committee asked Parker to make an exhibition jump, for which she was paid $30 of the $80 offered as prize money for that event.

The big event of the meet was the 50-mile free-for-all, for which a $1000 first prize had been offered. It was a closed-course race, and the women were anxious to prove that they could do well over a demanding course that required frequent high-speed turns around pylons. While rounding a pylon in her Waco biplane, however, Frances Marsalis failed to level out after making the turn. A wingtip struck the ground, and the aircraft crashed, killing Marsalis.

The race continued, and the initial results were posted as follows:

50-Mile Free-For-All Handicap Race		5 August 1934	
Pilot	Place	Time(min.)	Prize($)
Edna Gardner	Waco Taperwing	38.48	1000
Helen Richey	Eaglerock	38.53	500
Arlene Davis	Stinson R	39.00	250
Gladys O'Donnell	Monocoupe	39.10	150
Helen McCloskey	Monocoupe	39.20	100

However, the referee, Captain Victor Strahm, cancelled the remaining events, an aerobatic contest and a barrier landing contest, because he and the Contest Committee felt the women were under too much strain following Frances Marsalis' death to do anymore flying.

Later that day, Ruth Barron Nason and Annette Gipson both filed protests that Edna Gardner had violated the rule that prohibited an overtaking aircraft from passing a slower aircraft on the inside; the slower aircraft was to be passed on the outside. Nason stated that Gardner, who was flying at 120 miles per hour, had pased her Waco F, which was flying at 107 miles per hour, on the inside of the turn at the home pylon.

The Contest Committee's subsequent investigation upheld Gipson and Nason's protests and disqualified Gardner, making Helen Richey the winner. Nason moved up to fifth place, having completed the race in 39.36 minutes. (See Appendix 1 for full results of the meet.)

Edna Gardner, the reported winner, had become interested in aviation in the mid-1920s when she was a nurse and one of her patients, a U.S. Navy pilot, sparked her interest in flying. She began flying lessons in 1926, soloed and received her license in 1927. She became a flight instructor in 1930 and in 1936 began and operated the New Orleans Air College at Shusham Airport, which continued until 1941. During that time she became the first woman to receive a Department of Commerce instructor rating.

In 1932, while she was in the Navy Nurse Corps stationed in Newport, Rhode Island, Gardner was selected to make a transatlantic flight sponsored by the American Nurses' Aviation Service for the purpose of supporting aviation medicine. Although there would be two male pilots, Gardner was still eager to go. The flight would be from New York to Rome, and Gardner was supposed to parachute out over Florence, Italy, as a tribute to Florence Nightingale. However, the Navy would not grant Gardner a leave of absence from her job, and she was replaced by Edna Newcomer. Newcomer and the two pilots in their Bellanca aircraft *The American Nurse* disappeared without a trace somewhere near the Azores (Ackermann-Blount, 1984:104).

Gardner also had hopes of becoming a commercial airline pilot. She first applied to Central Airlines because they had just hired Helen Richey. By the time Gardner applied, however, Richey had resigned, and Gardner was not hired.

She next applied to Chicago and Southern Air Lines in 1935, which was hiring some of her students, but she was told she was too short. The last airline to which she applied in 1938 told her outright that they believed a woman in the cockpit would be bad for business (Ackermann-Blount, 1984:102).

Gardner continued with her flying school. Throughout the 1930s she took part in numerous women's air meets and air races, and also flew gliders and autogiros.

In the first year, 1935, that women competed in the Bendix Trophy Race, there were two female entries, Amelia Earhart and Jacqueline Cochran, the best known women in aviation in the 1930s and perhaps to this day.

Jacqueline Cochran had been born into poverty in northern Florida, but at a very early age had determined not to stay there. Using her skills as a hairdresser and beautician, she went to New York City in the early 1930s and took a job in a prestigeous New York salon. In 1932, to promote the cosmetics and perfume business she had started, Jacqueline Cochran, Inc., she took up flying. Cochran received her pilot's license after only three weeks instruction at Roosevelt Field, Long Island, in a Fleet trainer. Flying, however, soon became her primary career. Believing that she needed further instruction, she went to San Diego in 1933, where she enrolled in the Ryan School of Aeronautics and took instruction in a Great Lakes Sport. While in California, she purchased a Travel Air and got her friend U.S. Navy Lieutenant Ted Marshall to give her advanced instruction in techniques of spot landings and night flying. She later took instruction

1929- OXX-6 TRAVEL AIR 2000

1929- PRIMARY GLIDER

1926- OX-5 SWALLOW T.P.

1932- J6-5 ARISTOCRAT

1933- PITCAIRN AUTO GYRO

1932- J-5 TAPERWING WACO

1946- NORTH AMERICAN SNJ

1963- 1-26 SCHWEIZER

23,000 HOURS in 50 YEARS of FLYING

952- AERONCA S7EC

1928- VIELE MONOCOUPE

1966- AERO COMMANDER 200

1951- BELL 47-H

1975- FLIGHT INSTRUCTING

EDNA GARDNER WHYTE, PRES.-CHIEF PILOT
Aero-Valley Airport, Inc.
RT. 2 AC 817-430-1905 ROANOKE, TEXAS 76262

FIGURE 39.—Scenes from the flying career of Edna Gardner Whyte. (S.I. photo 76-2186)

FIGURE 40.—Edna Newcomer, who disappeared in 1932 in the Bellanca aircraft *The American Nurse*. (Photo from Rudy Arnold Collection, National Air and Space Museum)

FIGURE 41.—Jacqueline Cochran (bottom, third from right) and members of her class at the Ryan School of Aeronautics in 1933. (Photo courtesy of Ryan Aeronautical Library)

FIGURE 42.—The Gee Bee Q.E.D., which Jacqueline Cochran and Wesley Smith flew in the 1934 MacRobertson Race. (Photo from Rudy Arnold Collection, National Air and Space Museum)

equivalent to a U.S. Navy course on ground and air work.

Cochran entered her first major air race in 1934, the MacRobertson Race from England to Australia, flying as copilot to Wesley Smith in the Granville Brothers Q.E.D. This aircraft had replaced the Northrop Gamma which Cochran and Smith had planned to fly; the Gamma crashed on its delivery flight to England.

Cochran and Smith did not fare very well with the Q.E.D., either. They were forced down in Rumania by engine trouble and had to withdraw from the race. Cochran's attempt to leave Rumania was potentially even more dangerous than her forced landing. In her autobiography, she gave this account of the adventure (Cochran, 1954:54–55).

Smith stayed on to get the plane in flying condition and to bring it back to the boat. I got to a store to buy a coat to cover my flying clothes and within a matter of hours, was on the Orient-Express for Paris. At the border of Hungary, in the middle of the night, my door was opened by officials and request was made for my passport. The officials could not speak English and I could speak nothing else, but it was clear something was wrong. They made signs that I would have to get off the train. I backed into the far corner of my compartment with a water bottle in my hand and prepared to fight it out physically, if necessary. Suddenly I thought of the Bucharest evening paper in my handbag with a front page picture of myself and the Air Minister. I showed it to them and then pointed to my flying clothes. The officials backed away and left me in peace. The next morning, I learned from the International Conductor that I answered the description of a jewel smuggler they were looking for and they were more than suspicious when my passport bore no permit to enter Rumania. I had not planned to enter Rumania, except for an airport stop, nor to leave by train.

The Gamma was rebuilt for the 1935 Bendix Race, but it still had problems, mainly an inexplicable tendency for the engine to overheat. To add to these problems, a heavy fog moved in just at the start of the race, which that year would be from Los Angeles to Cleveland. Cochran's determination to continue was based on two factors: reluctance to allow her male competitors to accuse her of being a fair-weather pilot and a desire to compete against the already famous Amelia Earhart, whom she had only recently met (Cochran, 1954:62).

In addition to the mechanical and weather problems Cochran encountered that early morning, she also had the misfortune to witness the death of a fellow pilot. Cecil Allen crashed in his Gee Bee on takeoff, and Cochran's start had to be delayed until the wreckage had been cleared.

From the moment she started down the runway for her takeoff at 4:22 A.M. on 31 August, Cochran had problems with the engine. It was barely giving her enough power to lift off, and it was quite some time before enough airspeed had built up for her to reach cruising altitude. Over the Arizona desert the engine began to overheat badly, and with the Grand Canyon looming ahead, she turned back to Kingman, Arizona, and landed.

When asked why she landed, she later wrote that she said, "I just got tired and quit." She went on to explain, "I saw no reason to give an alibi which would put blame on the plane or the engine. The manufacturers involved had done their best to stop my flight" (Cochran, 1954:65). Indeed, because of the trouble with the aircraft, Northrop and Pratt & Whitney officials had tried to convince her to withdraw before the start, but Cochran felt it was essential that she take off and at least make an attempt at winning (Cochran, 1954:62).

Cochran's intensity was in sharp contrast to the relaxed atmosphere in Amelia Earhart's Lockheed Vega. At 12:34 A.M., Earhart saw enough of a break in the fog to take off. In the Vega's small passenger area, Earhart's advisor Paul Mantz and engine designer Al Menasco were playing gin rummy. Earhart knew her Vega was outclassed by Ben O. Howard's *Mr. Mulligan* and Roscoe Turner's Wedell-Williams racer. Indeed, although she had been the first to take off, Earhart was the last to land in Cleveland, 13 hours and 47 minutes after her takeoff. Howard's winning time was 8 hours and 33 minutes with an average speed of 238.704 miles per hour, compared to Earhart's 149.578 miles per hour. Her fifth place finish earned her $500, however.

During 1936, Cochran continued practicing her flying techniques and using an airplane to promote her growing cosmetics business. That year also she married Floyd Odlum, a wealthy aviation entrepreneur and California rancher.

Jacqueline Cochran returned for another try in the Bendix Race in 1937 as the only woman competitor. This time she was flying a green and white Beechcraft D-17W Staggerwing. She no doubt felt much more confidence in her Staggerwing than she had in her problem-plagued Gamma, because earlier that year she had set a new women's national speed record of 203.895 miles per hour in the Beechcraft.

Her determination to do well sprang from her intensely competitive nature, which had been strengthened by her disappointments in the 1935 and 1936 Bendix Races. There was another equally important factor that motivated Cochran that year. Following the tragic disappearance of her friend

FIGURE 44.—Cochran with the Beechcraft Staggerwing in which she finished third in the 1937 Bendix Trophy Race. (Courtesy of H. Glenn Buffington)

FIGURE 43.—Paul Mantz and Amelia Earhart just before taking off in the 1935 Bendix Race. (S.I. photo A38749F)

FIGURE 45.—Vincent Bendix congratulates Cochran after her victory in 1938 Bendix Trophy Race. (S.I. photo 79-3161)

FIGURE 46.—Alexander de Seversky and Cochran confer before the 1939 Bendix Trophy Race. (Courtesy of Fairchild Industries)

Amelia Earhart that July, Cochran shouldered the burden as standard-bearer for women pilots. In her keynote speech at the Women's National Aeronautical Association's ceremony honoring Earhart, Cochran acknowledged that Earhart had "placed the torch in the hands of others to carry on to the next goal, and from there on and forever" (Moolman, 1981:132).

At 12:04 A.M. on 4 September, Cochran was the first of the 15 competitors to take off in the Bendix Race. Her flying time to Cleveland of 10 hours and 29 minutes won her third place behind Frank Fuller in a Seversky and Earl Ortman in a Marcoux-Bromberg. This made her all the more determined to compete again in 1938.

Two factors contributed to Cochran's confidence in the forthcoming Bendix Race. In the 1937 race, two sleek new Seversky P-35s had been entered, the first military-type aircraft to compete in the Bendix. One of the problems women pilots constantly faced was the attitude that they could not handle "hot" aircraft built for speed, and therefore found their aircraft outclassed in the races. No matter if a woman's skills as a pilot were just as good as those of a male competitor; a stock Lockheed Vega had very little chance against a streamlined racer especially designed for the Bendix. When Cochran saw the P-35s as she waited to take off in her Staggerwing, she was determined that someday she would fly one.

In the meantime, Alexander P. de Seversky, designer and builder of the P-35, was having a great deal of difficulty convincing the Army Air Corps of the merits of the aircraft, even after Frank Fuller had won the 1937 Bendix in a P-35 in the record time of 7 hours, 54 minutes, 26 seconds at an average speed of 258.2 miles per hour. To further prove his product, Seversky added extra fuel cells in the wings to increase the aircraft's range, and he began breaking long-distance speed records. The Army Air Corps credited this, however, to Seversky's skill as a pilot rather than to the performance capabilities of the P-35.

So, in keeping with the adage, "If a woman can fly it, anyone can," Seversky turned to Jacqueline Cochran, who by that time had established herself as the premier female pilot in the United States. On 21 September 1937, she set a new world speed record for women of 292.271 miles per hour in Seversky's aircraft. Still the Air Corps was unmoved.

Seversky next offered Cochran a P-35 with extra tanks and a 1200-horsepower engine for her to use in the 1938 Bendix Race. No one had yet really tested this latest modification over a long distance. Cochran, the only woman in a field of 10, took off from Burbank early on the morning of 3 September. Over the Continental Divide, the engine quit. She switched tanks, but it became apparent that although there was enough fuel, it just was not getting to the engine. She rocked the wings and found that the engine ran quite smoothly with the left wing down. It was later found that a wad of paper had inadvertently been left in the right wing tank when the extra tanks were installed, and it had blocked the flow of fuel. Undaunted, Cochran won the 1938 Bendix Race, flying the 2042 miles to Cleveland in 8 hours and 10 minutes at an average speed of 249.774 miles per hour.

A new prize had been added in 1937 for the first Bendix Race pilot to fly from Los Angeles to Bendix Field, New Jersey, so after refueling at Cleveland, Cochran took off for the East Coast. She did not stop after reaching Bendix Field, however, but continued to Floyd Bennett Field, New York, establishing a new women's west-to-east transcontinental speed record of 10 hours, 7 minutes, and 10 seconds. Her winnings that day totaled $12,500—$9000 for the Bendix win, $2500 as the first woman in the race to reach Cleveland, and $1000 as the first pilot in the race to reach Bendix Field.

Cochran returned once again in 1939 with another Seversky P-35, intent on becoming the first two-time winner in the history of the Bendix Race. She picked up the aircraft at the factory on Long Island, and for its maiden flight headed west for the starting point of the Bendix in Burbank. She was forced to put down at Dayton, Ohio, with landing gear problems, and she also experienced some difficulty with the new, thinner wing with which that modification was equipped. As a consequence, Cochran was delayed in arriving at the Bendix starting point, and by the time she was ready to take off, the ceiling was down to 800 feet. Faced with withdrawing or taking a grave risk, Cochran withdrew.

When asked later why she usually tried for absolute records rather than for the women's records, Cochran replied that she did not want to end up holding only women's records because they were invariably broken by men in higher performance aircraft. She said of the difference between male and female pilots, "If they [women] know how to fly, they know how to fly, and they're not any better and not any worse" (Cochran, 1979). For Cochran, the key to success was dedication and endurance. She was once quoted as saying, "I think it [flying] takes courage . . . . But the thing we must remember is that courage, to be important, must endure. Anybody can take anything for a little while. But when you're blazing new trails, you have to be able to endure."[3]

Nineteen thirty-nine was an important year for Cochran. On 24 March, she set a women's national altitude record over Palm Springs, California, of 30,052.43 feet in a Beechcraft Staggerwing, and on 15 September she set an international women's speed record of 305.926 mph over a 1000-kilometer straight course between Burbank and San Francisco in a Seversky Pursuit. On 8 August, at the Pittsburgh airport, she made the first blind landing (solely by reference to instruments) by a woman pilot. By the beginning of World War II, Cochran was the country's leading female pilot. She had this to say some years later on the subject of male pilots' attitudes to her flying (Cochran, 1979).

Pilots respect other good pilots. I never found any beef or any lack of cooperation or anything with the opposite sex. I think that they hated men and women that pretended they flew and put on a big act, but if they really got out and did it, I think they [male pilots] respected them [female pilots]. I never had any problems, any.

There was another female competitor in the 1939 Bendix Race. Arlene Davis of Cleveland, with Dale Myers as passenger, finished fifth in a Spartan 7 aircraft to win $1000. She was disqualified from accepting the $2500 women's prize, however, because Myers was a licensed pilot, and Davis could not prove she was at the controls for the entire flight (Dwiggins, 1965:104).

Although they were not allowed to compete in major races, such as the Bendix, until the mid-1930s, women had been allowed earlier in that decade to take part in the National Air Races in events reserved exclusively for women. However, their times and speeds were very close to those of their male counterparts in comparable races. For instance, in the 1930 race for Women's Cabin Ships with 800-cubic-inch engine displacement (Event No. 35), Phoebe Omlie flying a Monocoupe had a winning time of 10 minutes, 43 seconds, at 139.97

miles per hour. Vern Roberts, also in a Monocoupe, won the same event for men in 10 minutes, 18 seconds at 145.58 miles per hour.

May Haizlip won the Women's 500-Cubic-Inch Open Ships event (No. 1) in an Inland Sport, flying the 5-mile, 5-lap course in 12 minutes, 23.9 seconds at a speed of 121.08 miles per hour. W.C. Moore, also flying an Inland Sport, won the Men's 450-Cubic-Inch Open Ships race in 11 minutes, 38 seconds at a speed of 128.85 miles per hour.

May Haizlip was well known in the 1930s as a record-setting air racer. In 1930, Haizlip flew a Cessna racer in the National Air Races, finishing second in the Women's Free-For-All. During the 1931 National Air Races, she competed in six different high performance aircraft including a Laird racer and the Gee Bee D, becoming the second highest money winner, male or female, at that year's races. During the 1932 National Air Races, Haizlip, in the Wedell-Williams #92, set a new speed record for women of 255.513 mph, a mark that stood until 1939, when it was broken by Jacqueline Cochran. In addition to her racing career, Haizlip served as a test pilot for Spartan Aircraft, American Eagle, and Buhl Aircraft.

Bettie Lund, another frequent air race competitor, became interested in aviation when she was about nine years old and won a ride in a Curtiss Jenny by selling magazine subscriptions. But it was not until after her marriage to the famous stunt pilot Freddie Lund that her own aviation career began.

Soon after the Lunds married in 1929, Bettie Lund began traveling to air meets with her husband. In his Waco Taperwing, painted in red, white, and blue barberpole stripes, Freddie Lund was well known in United States aviation shows for his daring aerobatics. Between performances, Freddie Lund taught his wife to fly. Because they traveled so much, Bettie Lund learned to fly in 13 hours at 13 airfields using 13 different types of airplanes, including her husband's Waco (Hyman, 1932).

Bettie Lund's first solo flight occurred in Miami in 1930 in a Curtiss Fledgling. Since all her flying lessons had taken place at busy airfields, usually during air meets, she had had virtually no taxi instruction, so she was told that after her first solo, she should simply roll to a stop and not move until her husband arrived.

After Bettie had made only two more solo flights, her husband decided she should try for her first record. She set a new women's barrel roll mark of

---

[3] Unidentified, untitled newspaper article by Adela Rogers St. John, 2 October 1938, Jacqueline Cochran biographical file, National Air and Space Museum archives.

FIGURE 47.—May Haizlip. (Photo from National Air and Space Museum archives)

FIGURE 48.—Freddie and Bettie Lund with his red, white and blue Waco Taperwing, Troy, Ohio, 1931. (Courtesy of Bettie Lund)

FIGURE 49.—Bettie Lund before her 1930 women's barrel roll record set in Miami. (Courtesy of Bettie Lund)

FIGURE 50.—1931 National Air Races, Cleveland, Ohio (left to right): Edna Newcomer, Marty Bowman, Gladys O'Donnell, Edith Foltz, Bettie Lund. (Courtesy of Bettie Lund)

FIGURE 51.—Gladys O'Donnell, winner of the 1930 Women's Pacific Derby, Long Beach to Chicago, in her Waco Taperwing. (Courtesy of H. Glenn Buffington)

67, and thus Freddie and Bettie Lund became one of the era's most famous husband and wife aviation teams. In addition to her aerobatics—of which she said her favorites were the barrel roll, slow roll, roll-and-a-half, and inverted flying (Driscoll, 1931)—Bettie Lund also took part in air races and dead stick landing contests.

In October 1932, Bettie Lund saw her husband crash to his death. She had had little time to realize what had happened when she received a call from the airport manager of Charlotte, North Carolina, inquiring about a contract Freddie had with them to appear at an air show. She answered, "I'll fly in his place" (Hyman, 1932). She flew two more air shows to fill her husband's contracts and then retired for a month to think things over. After that month, she bought a Waco Taperwing, painted it in barber pole stripes identical to Freddie's, and launched her own aerobatic career. By the end of the 1930s, she was one of the country's top stunt pilots. When asked why she continued to fly after her husband's death, she replied that it was not just her own love of aviation; it was a crusade to prove that aviation is safe (Hyman, 1932).

The 1931 National Air Races marked the dedication of the air race center at Cleveland. The races began with the Transcontinental Handicap Air Derby from Santa Monica to Cleveland. Sixty planes took off in the race, the largest group ever to compete in one event. Ten women were among the entrants, and to uphold the rule that women and men could not compete in the same contest, two separate divisions were created. The two divisions would fly the same course, and on the basis of the race's point system, one overall winner would be determined. The overall winner of the Derby was Phoebe Omlie with 109.19 points, for which she won a Cord Cabriolet automobile.

By 1930, Phoebe Omlie, a former barnstormer and movie stunt pilot, had established a respected name for herself in the aviation industry as an officer of Mid-South Airways and of the Mono-Aircraft Corporation. Omlie dedicated her life in aviation to proving the efficiency and versatility of the airplane. She used her abilities as a cross-country racing pilot to promote her ideas.

In 1930, Omlie led from start to finish in the Dixie Derby, a race, for women pilots, from Washington, D.C., through the southern states and up to Chicago. But in the 1931 Transcontinental Handicap Derby, Omlie was faced with her greatest racing challenge. She described later why she was so pleased to enter her Monocoupe in the race (Williams, 1932:15–16).

After looking over the rules and regulations of the 1931 National Air Derby, I found a new opportunity to try to prove to the world the efficiency and possibilities of the airplane I have devoted my time to since its entry in the aviation industry some five years ago.

In the two preceding derbies, which we won, we were fortunate in having the fastest airplane in the race. This made it possible for us to ease down on the throttle and cruise most of the way. But here, with a handicap derby, there was an opportunity to try and show the aviation and lay world that the equipment could be pushed to the utmost, and because of its efficiency, along with its speed and endurance, have a chance to win the Sweepstakes.

Omlie told reporters that she was pleased with her overall first place "because I had again been able to prove that the little ship was able to be first, not only in speed, but in efficiency, the thing that means so much to every airplane owner" (Williams, 1932:31).

In 1932, Omlie flew over 20,000 miles around the United States campaigning for Franklin D. Roosevelt for President. Her skills as a pilot and campaigner did not go unnoticed by F.D.R. In November 1933, he appointed her Special Advisor for Air Intelligence to the National Advisory Committee for

FIGURE 52.—Phoebe Omlie christens the official plane of the National Air Race Corp. at Curtiss-Wright-Reynolds Airport, Chicago, 23 June 1930. Captain Max M. Corpening, executive director of the races, and Major R.W. Schroeder, director of contests, are at right. (S.I. photo 80-19981)

FIGURE 53.—Phoebe Omlie (left) and Stella Aikin with the Democratic National Committee's plane at Floyd Bennett Field, New York, 16 September 1936. (Photo from Rudy Arnold Collection, National Air and Space Museum)

Aeronautics (NACA). Omlie instigated, planned, and directed the National Air Marking Program, beginning in late 1934. She continued as head of the program until it was well on its way to being a success, when she returned to her duties at NACA.

When her husband, Vernon Omlie, was killed in the crash of a commercial airliner near St. Louis in 1937, Phoebe resigned from government service to devote her time to research and development of air safety and flight training. She returned to her home in Memphis, where she authored the Tennessee Aviation Act of 1937. This was the first program to provide aviation training through the public schools, and it preceded by two years the U.S. Government's Civilian Pilot Training Program.

As part of the Tennessee program, Omlie directed and instructed the first class in Memphis of the Tennessee Civilian Pilot Training Program (TCPT) in 1938. She stressed to the school board the importance of making the program part of the regular curriculum. Omlie believed that early instruction in flying and respect for the airplane would lead to increased aviation safety, and she continued for many years in this pursuit.

The chairman for the women's division of the 1931 Transcontinental Handicap Air Derby was Florence Lowe Barnes. Few people, even aviation buffs, would recognize that name, but the name "Pancho" Barnes is legendary in aviation history. Barnes was the granddaughter of the famous 19th century balloonist Thaddeus S.C. Lowe, who during the Civil War was head of the Union Army's Aeronautic Corps. She got the nickname "Pancho," according to popular legend, while serving as a crew member on a banana boat that was running guns to Mexican revolutionaries.

Her work for Union Oil in the late 1920s as an air racer, stunt pilot, and aviation promoter led her to a job as a stunt pilot in the famous Howard Hughes movie "Hell's Angels." But Barnes' aviation career was not all stunts and racing.

On 5 August 1930, she established a new world speed record for women of 196.16 miles per hour over a measured course near Los Angeles in her Travel Air *Mystery Ship*. Earlier that year she had made the first flight by a woman from Los Angeles to Mexico City. Barnes set a Los Angeles–Sacramento round-trip speed record on 1 March 1931 and was presented a trophy by the Governor of California inscribed to "America's fastest woman flyer."

Barnes also used her organizational skills to help form a group of women pilots who were available to assist people during disasters, such as floods or snow storms. During a test flight near Big Bear Lake in California, Barnes dropped a crate of eggs 7,000 feet by parachute without breaking any (Author unknown, 1975).

During the 1930s, Barnes continued to participate in air meets and air races, but she devoted most of her time to the civilian pilot training school that she ran on her ranch in California's Antelope Valley. This became the famous guest ranch known as the "Happy Bottom Riding Club."

The 1931 National Air Races also saw marked increases in speeds by both male and female pilots. For instance, Maud Tait won the Women's Free-For-All (Event No. 33) in a Gee Bee Y at 187.57 miles per hour, 38 miles per hour faster than Gladys O'Donnell's winning time the year before. During the races attempts were made to set new landplane speed records. Maud Tait in her Gee Bee Y set a new international women's record of 210 miles per hour.

The program for the 1932 National Air Races was the first one that did not carry the statement, "Men and women pilots will not be allowed to compete in the same events." In the 19 races in which women were allowed to compete with men, women placed in the top five in six of them. May Haizlip's speed of 255.51 mph in the Shell Speed Dash (Event No. 29) set a new world speed record for women.

In 1933, there were only two events in which women could participate, and one of those, the Women's Shell Speed Dash, was not flown. Instead, the $1425 in prize money was awarded based on the results of the Aerol Trophy Race.

After a year of holding their own race meet, women returned to the National Air Races in 1935. Their results were impressive and are reported in the Appendixes. One of the most coveted trophies for which the women competed was one that bore Amelia Earhart's name (Event No. 23). When the trophy was first awarded in 1932, the distance was 21 miles, 6 laps over a 3½-mile course. The first winner was Florence Klingensmith in a Monocoupe. The contest was not held in 1933 or 1934; in 1935, it was won by Edith Berenson in a Bird. By 1937, the course distance had increased to 25 miles, 5 laps over a 5-mile course. Eligible aircraft had to have an engine displacement of 800 cubic inches or less and a maximum speed of under 175 miles per hour. Contestants had to qualify at a speed of

FIGURE 54.—"Pancho" Barnes and her "Mystery Ship."
(S.I. photo 74-3400)

FIGURE 55.—"Pancho" Barnes. (Courtesy of Edwards Air Force Base
History Office)

FIGURE 56.—Earhart awards the 1932 Amelia Earhart Trophy to Florence Klingensmith.
(S.I. photo 84-6307)

100 miles per hour or better. Also in 1937, the race's name was changed to the Amelia Earhart Memorial Trophy Race.

The number of races that made up the National Air Races in 1938 and 1939 had been drastically reduced from what it was in the early 1930s. In those two years, the only race in which women competed was the Bendix Race. But the National Air Races was certainly not the only racing event in which women pilots took part.

Not exactly a race but a contest for sportsmen pilots was the Ruth Chatterton Air Sportsman Pilot Trophy Race, held in conjunction with the National Air Races beginning in 1935. The Derby was not a race to see which pilot could fly the legs of the route in the fastest time, but a test of precision flying, with the winners being the flyers who could navigate and pilot their aircraft the most accurately.

The idea for the race was conceived by Cliff Henderson, who wanted to involve sport pilots in the National Air Races. Ruth Chatterton, an actress who was also a private pilot, agreed to sponsor the contest, for which a total of $5000 in prize money would be awarded. The transcontinental course was determined, and the lap distance between airports was measured and provided to each contestant. A control aircraft flying at a predetermined speed established the par time for each lap and determined corrections that might have to be made for adverse weather conditions. Each pilot named his or her own handicap in relation to the control plane, and pilots could change their handicap before taking off from any control point.

A typical route for the Chatterton Air Derby was the one flown in 1936 from Cleveland to Los Angeles, the site of that year's National Air Races. Takeoff from Cleveland was on 29 August, and the finish would be 4 September, the opening day of the races.

FIGURE 57.—Edith Foltz stands beside the Alexander Bullet in which she completed in the light plane class of the 1931 Women's National Derby from Santa Monica to Cleveland. Judges refused to allow her to take her dog because of the ironclad rule of the race, "No males allowed." (S.I. photo A5025D)

FIGURE 58.—Katherine Cheung. (Courtesy of Katherine Cheung)

FIGURE 59.—Babe Weyant winning the 1940 Joyce Hartung Trophy for the Michigan female pilot who accumulated the most points at an annual Michigan aeronautical event. (Courtesy of Babe Weyant Ruth)

The route that year was Cleveland, Cincinnati, Louisville, Nashville, Memphis, Hot Springs, Dallas, Big Spring, El Paso, Douglas, Tucson, Yuma, San Diego, Long Beach, and Los Angeles. The Chatterton Race was held only in 1935 and 1936 (the results for both are given in Appendix 1).

One of the participants in the 1936 Chatterton race was Katherine Cheung, the first woman pilot of Chinese ancestry to earn a pilot's license in the United States. Cheung was a native of Canton, China, but had moved to the United States to join her parents in 1927.

In 1931, Cheung was an accomplished pianist studying at the University of Southern California when she heard from some friends that the flying schools in China were not admitting women. Apparently this was not because Chinese authorities objected to women taking flying lessons, but because there were so many men who wanted instruction that the schools were full.

Cheung decided to take up flying herself and return to China to open a flying school for women only. Pursuant to this goal, she earned her license and studied aerobatics and blind flying. In 1937, she began flying around to various Chinese communities on the West Coast in her Fleet aircraft, raising money for her school. By the next year, she had $7000 and with part of that she bought a Ryan ST trainer. She eagerly awaited delivery in San Diego, but her dream was shattered. Her mechanic, who was checking out the aircraft, allowed a male Chinese student pilot to take the aircraft up. The young man crashed the aircraft while coming in for a landing; Cheung never flew again.

A common event in the 1920s and and 1930s was the air tour, another device intended to make Americans more "air-minded." The Michigan Air Tour, sponsored by the Michigan American Legion, was typical of these events. The 1932 Air Tour began in Detroit and made 16 stops in Michigan cities. Pilots were entertained at luncheons, dinners, and teas. In turn they reciprocated the host towns' hospitality by taking local residents for rides at a cost of $1.50 each. In the 1932 tour were three Michigan pilots: Vera Brown of Detroit, Ruth Kitchell of Coldwater, and Wilfred Blair of Grand Rapids.

One woman who was particularly active in the Michigan Air Tour was Marion ("Babe") Weyant. Weyant had wanted to learn to fly in 1931, but she was thwarted because of lack of funds and because she was only 13! The latter problem could only be solved by time, but Weyant immediately went to work on the former. She began by raising chickens for profit, and in 1934 was allowed to operate a candy and soft drink stand at the Lansing airport. She soloed in 1936, and in the 1937 Michigan Air Tour, Weyant, the youngest pilot in the 53-plane field, was chosen to lead off the group in her Aeronca. Weyant also took part in the 1938 and 1939 Air Tours, and in the latter event, she took along as a passenger the man who had initially opposed her desire to learn to fly—her father.

A Michigan Girls' Air Day was held 25 September 1938, at Hartung Airport in Detroit and attracted women pilots from all over the state to compete in such events as ribbon cutting, spot landing, and bomb dropping. Occasions such as these were designed to bring women pilots together and to show the attendees what women could do in aviation.

There were many other events held around the United States in the 1930s that were designed to heighten peoples's awareness of aviation. One of the best known of these was the Ford Reliability Tour, which had begun in the late 1920s with the purpose of proving that airplanes could keep a schedule and could be a reliable means of transportation. The only woman in the 1930 Reliability Tour, in a field of 35, was Nancy Hopkins of Long Island. Hopkins had engine trouble over the Mississippi River but landed safely in a field in Arkansas, repaired the problem herself, and managed to arrive at her scheduled stop on time.

One of the more unusual aviation events of the 1930s was the Liberty Aerial Treasure Hunt, held in conjunction with the National Charity Air Pageant at Roosevelt Field, 7 and 8 October 1933. Bernarr Macfadden, editor of *Liberty* magazine, was the sponsor of the treasure hunt, and the prize money totaled $1918. Beginning in St. Louis, contestants received a clue in verse at each stop. Clues would lead them to the next city, and in each city the pilots would look for a white cloth 50 feet by 50 feet with a letter of the word "LIBERTY" on it. The pilot would then indicate on a map where each letter was found. The first pilot to arrive at Roosevelt Field with his or her map correctly marked was the winner.

A typical clue was the one the pilots were given in St. Louis (Author unknown, 1933:25).

> Leaving St. Louis straight to rear,
> O'er a SOCIAL MOOD you steer.

And down below you will see
If you read as well as me,
The first name of an Engineer.

The "straight to the rear" would tell the pilot to cross Illinois, the "SOCIAL MOOD" would suggest the Embarrass River, and in the town of Casey (for the engineer Casey Jones) the pilot would find a large "L" and the next clue.

Five women were entered in a field of 21: Helen McCloskey, who finished in second place; Helen Richey; Peggy Remey; Genevieve Savage; and Laura Jackson.

By the outbreak of World War II, much had changed for aviation in general and for women

pilots in particular. Air travel was a much more commonplace mode of transportation. Manufacturers and aircraft sales companies no longer felt the need to continue hiring women as demonstration pilots and saleswomen, because most of the novelty—and the fear—was gone from private flying. The glamour and excitement of the air races had dimmed somewhat, so that women could no longer count on winning races as a means to making names for themselves in aviation.

By making aviation tamer and more acceptable, women had, ironically, closed several doors to paths they had previously followed toward fame in aviation.

Courage That Endures

Among the many women who flew in the 1930s were several who found themselves in the limelight more often than the others. They became well known to the American public (and in some cases, to the world) through their record flights.

These were women for whom aviation was more than a sport, a pastime, or a passing fancy. Aviation was their career, almost their *raison d'être.* When Louise Thaden was grounded before the birth of one of her children, she had these thoughts on her longing to be up in the air (Thaden, 1938:94–95).

As much as anything else, I missed the soothing splendor of flight. The ability to go up into God's heaven, to look out toward distant horizons, to gaze down upon the struggling creatures far below, to forget troubles which so short a time before seemed staggering, just to feel the lifting of the wheels from the ground, to hear the rush of air past the cabin window, to squint into the sun, toying with the controls, to feel the exhilaration of power under taut leash, responsive to whim or fancy, to feel, if only for one brief moment, that I could be master of my fate—that is what I missed!

The courage and tenacity of these and other female flyers of that decade did indeed endure and was an inspiration for younger women who also dreamed of flying. The careers of several of the women are described in this chapter.

A former vaudeville dancer who was once told she was too short to fly became one of the leading pilots of the 1930s. Laura Ingalls began the decade with setting two rather unusual records. On 8 May 1930, at St. Louis' Lambert Field, Ingalls set a new women's record for consecutive loops by executing 344 of them and, thereby, smashing the previous

record of 46, which had been held by Mildred Kauffman. Ingalls' record, which had taken 1 hour and 3 minutes, was set in her de Havilland Gypsy Moth at a height of about 8000 feet. Later that year, on 14 August, again in St. Louis, Ingalls performed 714 barrel rolls, breaking Dale Jackson's record for men of 417 barrel rolls and Bettie Lund's women's record of 67. The feat took 3 hours and 39 minutes. Ingalls said she landed then because "it began to get dark and I thought my wings were getting wobbly" (Author unknown, 1930d).

People were intrigued by Ingalls' records; aviation enthusiasts in Muskogee, Oklahoma, invited Ingalls to Muskogee and offered to pay her $1.00 for every loop she could do over her record of 344. They may have later regretted the offer, because Ingalls executed 930 loops, costing her hosts $636 (Author unknown, 1934:32).

Ingalls did not confine her records to loops and barrel rolls. She held more women's U.S. transcontinental air records than any other female pilot in the 1930s. Her first one was set during the Women's Transcontinental Derby of October 1930, when she made a round-trip flight between New York and Los Angeles, 30 hours east to west, 25 hours west to east.

The most famous flight of her career occurred in the spring of 1934, setting many new records: first flight over the Andes by an American woman; longest distance flight by a woman, 17,000 miles; first flight by a woman from North America to South America; first solo flight around South America; first flight around South America in a landplane

48

FIGURE 60.—Laura Ingalls. (S.I. photo A45972)

FIGURE 61.—Laura Ingalls in the cockpit of *Auto-da-fé*, 10 July 1935, before taking off on her record flight from Floyd Bennett Field to California. (Photo from Rudy Arnold Collection, National Air and Space Museum)

FIGURE 62.—Laura Ingalls atop the fuselage of her Lockheed Air Express at Floyd Bennett Field, New York, in 1932. (Photo from Rudy Arnold Collection, National Air and Space Museum)

(Author unknown, 1934:32). The aircraft was a Lockheed Orion.

Ingalls left Miami on 8 March and made stops in Havana; Mérida, Mexico; Managua; Cristobal, Canal Zone; Talara and Lima, Peru; Arica, Antofagasta, and Santiago, Chile. From Santiago, she crossed the Andes to Buenos Aires, then northward along the coast to Rio de Janeiro. Her return flight to the United States took her to Trinidad, the West Indies, Cuba, and back to Miami. She returned to Floyd Bennett Field, New York, on 25 April, three months to the day after her take off from there. For this flight, Ingalls was awarded the Harmon Trophy for the outstanding female aviator of 1934. She did not rest on her laurals, however. 1935 was Ingalls' big year for transcontinental records. In July, she made the first nonstop east-to-west flight by a woman, flying from Floyd Bennett Field, New York, to Union Air Terminal, Burbank, California, in 18 hours, 23 minutes. This established a new women's record and also bettered the east-to-west men's record by 5½ hours. Her aircraft was a Lockheed Orion she had named *Auto-da-fé* (Act of Faith).

On 12 September 1935, Ingalls flew nonstop from Los Angeles to New York in 13 hours, 34 minutes, breaking the women's record of 17 hours, 7 minutes, which was held by Amelia Earhart. Ingalls described long-distance flying to a reporter this way: "There's no thought of danger. It's a game. You watch for your landmarks, measure your time, smile, go on. Every real flier becomes one with the elements and just hates to land."[4]

Another woman of enduring courage was Ruth Nichols, who by 1930 had already established a name for herself in aviation as one of the founders of *Sportsman Pilot* magazine and of the then-popular aviation country clubs. She was one of the most outspoken women of that era on the ideas that flying is safe and that women make good pilots. Nichols devoted her career in aviation to proving those two premises. After one of her many record flights, she wrote the following (Nichols, 1932:80).

. . . the ideas and designs which were evolved can eventually be incorporated into standard equipment for the commercial transport lines. But the chief point of interest is the fact that these ideas could not have been developed had not the flights been of a spectacular enough nature to produce news value. News is a salable asset and has a concrete value in dollars and cents, which can beget money with which to carry on experiments.

---

[4] "Time For Love, Would Lift Her Sisters from the Kitchen to a Cloud," unidentified, undated newspaper clipping, Laura Ingalls biographical file, National Air and Space Museum archives.

Another reason for such flights, particularly by women, is to stimulate further public confidence in the safety and advanced stage of aviation.

Nichols was the only woman to hold simultaneously the women's world speed, altitude, and distance records for heavy landplanes. From 24 November to 1 December 1930, Nichols set a U.S. transcontinental speed record for women from Mineola, New York, to Burbank in 16 hours, 59 minutes flying time. On her return flight she was the first woman to make a one-stop transcontinental flight, flying from Burbank, Los Angeles, to New York in 13 hours, 21 minutes, breaking Charles Lindbergh's record of 14 hours, 23 minutes, set earlier that year on 20 April. On 6 March 1931, she set a world altitude record for women of 28,743 feet, breaking Elinor Smith's previous record by 1325 feet, even though her wing-mounted oxygen equipment froze. On 13 April 1931, she set a women's speed record of 210.704 miles per hour, and on 24–25 October 1931, she set a women's nonstop distance record of 1977.6 miles, from Oakland, California, to Louisville, Kentucky.

The October 1931 flight that set a distance record was made while her back was in a steel brace. In the summer of 1931, she had flown to St. John's, Newfoundland, on the first leg of a transatlantic attempt. She found the airfield there in very bad condition, but because she was low on fuel, she had to land. She overshot the runway, went around again, failed to clear the nearby hills, and crashed, cracking five vertebrae.

The aircraft Nichols used for all these record flights was a Lockheed Vega supplied to her by the Crosley Radio Corporation. Nichols and her technical advisor Clarence Chamberlin wanted an aircraft that would put to rest the notion that one aircraft alone could not serve for speed, altitude, and distance flights (Nichols, 1932:83).

In airplane building, one of the accepted fundamentals has been that a ship designed to shatter speed records should have short wings and be "all engine"; that a ship intended for altitude records should have large wings to give extra lift; and that a ship for long distance flight records should have a large body for installation of extra gas tanks. Three different designs, each highly specialized. My technical adviser, Colonel Clarence Chamberlin, and I figured that all these characteristics were necessary for an airplane which could make the transatlantic crossing with maximum safety. So we set to work to develop such a plane.

However, before Nichols could finish modifications to the Vega, Amelia Earhart became the first woman to fly nonstop, solo across the Atlantic in May 1932. Nichols continued to prove the versatility

FIGURE 63.—Ruth Nichols with the Lockheed Vega in which she set a world altitude record for women in 1931. The aircraft was used to set several other records in the 1930s and was the official radio plane of the 1931 Ford Reliability Tour. (S.I. photo 79-3165)

FIGURE 64.—Ruth Nichols in high altitude flight gear with her technical advisor Clarence Chamberlin. They are holding the barograph that confirmed Nichols' record altitude. (S.I. photo 75-4504)

FIGURE 65.—Elinor Smith and an official inspect the barograph following her 1930 record altitude flight. (S.I. photo 79-5962)

FIGURE 66.—In August 1930, Elinor Smith purchased a Lockheed Vega, which she named *Mrs. ?*, in which she hoped to fly nonstop from New York to Rome. Smith did not make the flight, and that same Vega was later acquired by Amelia Earhart for her 1935 Pacific flight. (Photo from Rudy Arnold Collection, National Air and Space Museum)

of her aircraft by setting an American altitude record for men or women in a diesel-powered aircraft, flying to an altitude of 19,928 feet in February 1932 at Floyd Bennett Field, New York.

Nichols also promoted the air transportation industry in several other capacities. In May 1932, she served as "Air Ambassadress" for the National Council of Women, the largest group of women's clubs in America. The purpose of this 3000-mile good-will tour was to educate women about the air transportation industry. Also in 1932, Nichols was asked to be a pilot on the first flight of New York and New England Airlines, a flight from New York City to Bristol, Connecticut. She went on to serve as air traffic manager of the airline, as well as being a reserve pilot.[5] Nichols spent the winter of 1933–1934 lecturing to organizations and colleges, talking to science departments about practical applications of aviation technology.

In the fall of 1939, with war breaking out in various parts of the world, the need for "mercy flying" was uppermost in her mind. Air ambulances could do great service in disaster relief and everyday civilian emergencies. At the annual aviation convention of the National Aeronautic Association in Washington, D.C., Nichols announced the founding of "Relief Wings," whose purpose was a humanitarian air service to provide relief, aero-medical information and facilities for civilian disasters of nature or war. Serving as its national director, Nichols, with three other women, set out on a 5-month national tour to raise funds and launch the program. This flight laid the foundation for Relief Wings in 36 states.[6]

Ruth Nichols used her fame and ability as a pilot not only for her own gain and women's integration into the industry, but to further the cause of aviation in general and serve society. Her vision was larger than most as shown by the admiration she expressed for the "unsung heroes" of aviation who test new equipment (Nichols, 1932:82).

Most of us women flyers have received far more attention than is our due, but there is a class of flyer of whom you should hear and whom you should remember. He is the pilot who carries modern flying beyond its safe modern technique, in order to find

a still safer one. And so, in future days, with every mile you fly in safety and comfort, you will realize the debt you owe to the present ballyhoo of aviation and to the great number of aviation mechanics and pilots whose names are unknown and whose feats are unheralded, who live quietly, work strenuously, and fly deftly, behind its raucous banners.

Elinor Smith and Bobbi Trout are two well-known women fliers of the 1930s who seemed to make a habit of breaking each other's records.

By 1930, 19-year-old Smith had made a name for herself in aviation with many altitude and endurance records. Her first in the new decade was achieved on 10 March 1930, when she set a women's world altitude record of 27,418 in a Bellanca Pacemaker at Roosevelt Field, Long Island. The feat almost cost her her life, as she passed out at the peak of the climb and recovered just as her aircraft was going into a dive. That record and the many others she had set in the 1920s won Smith the honor of being named one of the three best pilots in America by the American Society for the Promotion of Aviation in September 1930. She was singled out as the best woman pilot; also honored were then-Major Jimmy Doolittle, for his combination of flying skill and engineering knowledge, and Eddie Brooks, who was called the safest commercial pilot then flying (Author unknown, 1930b). Because she was a woman, however, the press also focused on Smith's flying attire. The 2 July 1930 edition of the *New York Sun* carried this account.

Girl Flyer Appears Wearing Real Shorts

Roosevelt Field's hard boiled aviators were goggle-eyed yesterday at the appearance of a flyer in warm weather "shorts." The latest advocate of the cool garb was Miss Elinor Smith, holder of the woman's altitude record.

Elinor wore a white, short-sleeve shirt, open at the neck, white slippers and socks. The shorts were of white duck and came within an inch of her knees.

When Ruth Nichols broke Smith's altitude record on 6 March 1931, Smith immediately set out to regain it. This time she used a supercharged Bellanca Skyrocket and took off from Roosevelt Field on 27 March 1931. Again, Smith passed out during the attempt, and she later gave this account to reporters (Author unknown, 1931c).

My motor cut out at 25,000 feet I tried to fool around and get it going. The oxygen tube fell out of my mouth and I became unconscious. When I came to I was going down. I tried to slide among the trees to make a good landing.

The right wing, the cowling, and the propeller of the Skyrocket were smashed, but the plane, and Smith's cuts and bruises, were soon mended, and

---

[5] "Biographical Sketch of Miss Ruth Nichols," unidentified, undated newspaper clipping, Ruth Nichols biographical file, National Air and Space Museum archives.

[6] "Biographical Sketch," unidentified, undated newspaper clipping, Ruth Nichols biographical file, National Air and Space Museum archives.

she was back in the air for another attempt on 9 April. This time she succeeded and appeared to have reset the women's world altitude record at 32,500 feet. Smith described the flight as follows (Author unknown, 1931d).

> The oxygen was so cold and dry that it took most of the skin off the top of my mouth. Due to the fact that my gas gauges were covered and, at that altitude, the effort to unhook the button covering them was so great, I soon experienced slight pains in the arms and chest, but as long as I sat still and breathed deeply of my oxygen I noticed that I could combat this.
>
> I have nothing but praise for the ship and motor; they both acted splendidly.
>
> It was unfortunate at the time of my previous attempt two weeks ago that people gained the mistaken idea that there was serious trouble with the motor, due to faulty construction. This is not so, as these motors for high-altitude work are so supercharged as to be entirely experimental. They are not standard jobs. Hence I consider the performance of the last attempt to be perfect from the performance standpoint.

With that flight, Smith not only would have taken the women's altitude record from Ruth Nichols but would also have beaten George Haldeman's record for men of 30,458 feet. However, later that month the Bureau of Standards informed the National Aeronautic Association that their calibrations of the barograph showed Smith only reached 24,951 feet, 3792 feet short of Nichols' record.

Smith was plagued that year with minor crashes and injuries, but in October she set a straight-course speed record for women of 229 miles per hour in a Consolidated Fleetster. In May 1932, Smith flew a Lockheed Vega from Los Angeles to New York on the first leg of what she hoped would be a nonstop flight from New York to Rome. The flight had to be abandoned, however, when the effects of the Depression forced her financial backers to withdraw. The Vega was later sold to Amelia Earhart, who used it for her 1935 Hawaii-to-Oakland flight.

Smith did not spend her entire aviation career in the air. From 1930 to 1935, she served as an aviation commentator for NBC radio, covering such events as the *Graf Zeppelin*'s landings in the United States, the flight of the *Southern Cross*, and the National Air Races. She also was instrumental in shaping aviation policy. In 1932, Smith was co-author of a New York State bill to ground uninsulated power lines near airports, and in 1936, she co-authored legislation to set up a separate division for aeronautical law in the New York State courts system. As an official advisor to the New York State Aviation Committee, Smith took part in a 1938

aerial survey on the possible sites of new airports in the state.[7]

Smith also served as aviation editor of *Liberty* magazine in the 1930s and was a frequent contributor to such journals as *Colliers, Popular Science, Aero Digest, Vanity Fair,* and *Cosmopolitan.* An article written about Smith in 1930 summed up her goals in aviation as follows: "She wants to convince manufacturers that women can become good pilots. She wants to make commercial aviation popular for women" (Wheeler, 1930:18).

Bobbi Trout started the 1930s in the same manner as she had ended the previous decade—by breaking a record. On 2 June 1930, she set a new women's cabin monoplane altitude record of 15,200 feet over Los Angeles. Endurance flights, however, were Trout's specialty. She and Elinor Smith exchanged the women's record three times before they finally teamed up in 1929 to set one together of 42 hours, 5 minutes.

In 1930, Trout and Edna May Cooper, an actress turned pilot, set their sights on the men's endurance record of over 645 hours held by Dale Jackson and Forest O'Brine of St. Louis. But their first attempt in their Curtiss Robin *Lady Rolph* on 2 January 1931 ended after only three hours because of a problem with the refueling plane's engine. They took off again two days later into the cloudy skies above Los Angeles. The two pilots had to combat not only boredom and fatigue, but also complaints by residents of Inglewood, California, that they were flying too low and making too much noise (Author unknown, 1931b). They also had to contend with such annoying incidents as the rope for their dinner basket becoming entangled in the plane's control wires. Cooper had to climb back along the fuselage to take care of the problem (Author unknown, 1931a).

Trout and Cooper's attempt at the men's record was cut short after 122 hours, 50 minutes, when a cracked piston caused smoke to fill the cockpit, and the flyers had to land. They had almost tripled the women's record, however. Eleven refueling contacts had been made during the five-day flight.

Trout then began planning what would be her first major distance flight since the 1929 Women's

[7]"Elinor Smith," unidentified, undated newspaper clipping, Elinor Smith biographical file, National Air and Space Museum archives.

FIGURE 67.—Edna May Cooper (left) and Bobbi Trout on New Year's Day 1930, with their Curtiss Robin *Lady Rolph*. The aircraft was named in honor of the wife of the Governor of California. (Courtesy of United Press International, Inc.)

FIGURE 68.—Bobbi Trout. (Courtesy of United Press International, Inc.)

Air Derby. She was going to attempt a Hawaii-to-U.S. flight in August 1932 in a Lockheed Altair. However, she could not gain any financial support and had to abandon her plan. Trout devoted the latter part of the 1930s to her work with the Women's Air Reserve, which helped in preparedness for national emergencies and disasters.

One women's altitude record that was set in the 1930s stood for three decades. When the Russian cosmonaut Valentina Tereshkova went into space in 1963, she finally took away from Jeannette Piccard the altitude record for women, a record Piccard had held since 1934. However, the record was set in a balloon and Piccard was first and foremost a scientist, not a pilot. She had worked with her husband Jean and his brother Auguste on readying the 600,000-cubic-foot balloon *A Century of Progress* for what turned out to be a world record altitude flight of 61,237 feet in 1933 with T.G.W. Settle and Chester Fordney aboard. The Piccards had been interested in using the balloon to study cosmic rays, and the main objective of the Settle and Fordney flight had turned out to be altitude.

The backers of the 1933 Chicago World's Fair, also known as the Century of Progress Exhibition, turned the balloon over to Jean and Jeannette Piccard, and they began preparations for a strato-

FIGURE 69.—Jean and Jeannette Piccard.
(Courtesy of Rev. Jeannette Piccard)

spheric flight. There were several problems, however, the first being that neither Jean nor Jeannette
Piccard had a balloon pilot's license. They decided
that Jeannette would get a license, and Jean would
serve as the scientific observer on the flight.

Jeannette Piccard began her flight training on 15
May 1934, with Ed Hill, the winner of the 1927
Gordon Bennett Balloon Race, as her instructor. She
soloed on 16 June. When she was asked later why
she had decided to become a pilot rather than simply
hiring one, she replied, "How much loyalty can you
count on from someone you hire?" (Maravelas,
1980:17). When asked if she also had parachute
training, she said, "No . . . if, on the first time you
jump, you don't succeed, there's no use trying again.
That was the training of the period" (Maravelas,
1980:17).

Another problem the Piccards faced was that of
financial backing. There was much opposition to
the fact that a woman would be the pilot. The
attitude of the National Geographic Society, a tra-
ditional backer of such ventures, was typical. "The
National Geographic would have nothing to do with
sending a woman—a *mother*—in a balloon into
danger," Jeannette Piccard later said (Maravelas,
1980:17). However, the Piccards did not share this
protective view. They finally were able to obtain

FIGURE 70.—Jeannette Piccard atop gondola just before takeoff,
23 October 1934. (Courtesy of Rev. Jeannette Piccard)

funds and assistance from several firms, such as
Grunow Radio and the People's Outfitting Company, and from private individuals; Henry Ford let
them use the Ford Airport at Dearborn, Michigan,
as their launch site (Crouch, 1983:629).

At 6:51 A.M. on 23 October 1934, the Piccards
lifted off. Jeannette was standing atop the gondola,
and Jean and their pet turtle were inside (Crouch,
1983:629). The balloon rose to a height of 57,579
feet. Eight hours after launch, the Piccards landed
safely near Cadiz, Ohio. With that flight, Jeannette
Piccard became the undisputed holder of the wom-
en's altitude record. Always eloquent, Piccard gave
this answer to the question of whether or not she
was afraid before making the flight.

Even if one were afraid to die there is so much of interest in a stratosphere trip that one does not have time to be afraid. It is too absorbing, too interesting . . . . When one is a mother, though, one does not risk life for a mere whim. One must back up emotions by cold reason. One must have a cause worthy of the danger. In times of war one sacrifices self and children for country. In times of peace the sacrifice is made for humanity. If one can forward by so much as even a little bit the sum total of man's knowledge one will not have lived in vain. If we do not add something to the knowledge of cosmic rays by our trip to the stratosphere this summer, we had better not go. We had better stay on the ground, be hewers of wood and drawers of water.[8]

---

[8] Jeannette Piccard, untitled, undated manuscript, "Speech and Writing File," box 77, Piccard Papers, Manuscript Division, Library of Congress.

Conclusion

When the United States entered World War II, women were allowed to use their aviation skills as part of the war effort. In 1942, Jacqueline Cochran took a group of American women to England to fly with the British Air Transport Auxiliary; one of those women was Helen Richey. Betty Gillies joined Nancy Harkness Love's small group of women in the Air Transport Command. Bobbi Trout owned two defense-related businesses.

FIGURE 71.—Helen Richey in her uniform of the British Air Transport Auxiliary during World War II. (Courtesy of H. Glenn Buffington)

Women were finally being taken more seriously as good, professional pilots. But at the same time, they had worked themselves out of the limelight and, in many cases, out of a job. Women in flying was no longer a novelty. The airlines did not need them to promote business, and aircraft manufacturers were wrapped up in the war effort; thus there was not the same market as that in which women had demonstrated and sold private aircraft. War clouds, not air show smoke, were in the skies, and the major air races, usual showcases for women pilots, were suspended.

Women were no longer oddities in any area of aviation. Be they stewardesses, engineers, businesswomen, or pilots, they had for the most part accomplished their goals of helping make air travel a standard means of transportation, and had proven to the world that women could be competent pilots.

The paths of succeeding women pilots, while still somewhat rocky, were smoothed in large measure by the enduring courage of women who were flying in the 1930s.

Appendix 1

Air Races of the 1930s in which Women Participated

1930 NATIONAL AIR RACES

Pilot	Plane	Time (h:m:s)	Prize ($)
Women's Pacific Derby—Long Beach to Chicago			
Gladys O'Donnell	Waco	15:13:16	3500
Mildred Morgan	Travelair	21:08:35	2100
Jean LaRene	American Eagle	21:45:49	1400
Ruth Stewart	Curtiss Robin	26:38:06	
Ruth Barron Nason	Buhl	38:33:41	
Margery Doig	Pitcairn	Forced out at Emporia, Kansas	
Women's Dixie Derby—Washington, D.C. to Chicago			
Phoebe Omlie	Monocoupe	11:42:21	2000
Marty Bowman	Fleet	14:48:39	1200
Laura Ingalls	De Havilland Moth	16:47:26	800
Nancy Hopkins	Kittyhawk	19:18:18	
Charity Langdon	Whitworth Avian	20:44:47	
Event No. 1: Women's 500-Cubic-Inch Open Ships			
May Haizlip	Inland Sport	0:12:23.9	500
Vera D. Walker	Inland Sport	0:12:25.9	300
Laura Ingalls	De Havilland Moth	0:13:16.8	200
Bettie Lund	Inland Sport	0:14:07.5	
Event No. 2: Women's 500-Cubic-Inch Cabin Ships			
Phoebe Omlie	Monocoupe	0:11:05.7	500
Marty Bowman	Monocoupe	0:11:11.6	300
Gladys O'Donnell	Monocoupe	0:11:22.3	200
Event No. 32A: Women's Free-For-All			
Gladys O'Donnell	Waco	0:20:00.8	1250
May Haizlip	Cessna	0:20:12.8	750
Opal Kunz	Travelair	0:20:34.9	500
Margery Doig	Pitcairn	0:21:32.5	
Phoebe Omlie	Monocoupe	0:22:17.6	
Ruth Nichols	Laird	Disqualified*	
Mildred Morgan	Travelair	Landed at 5th lap	
* Cut #1 pylon, 3rd lap.			
Event No. 34: Women's 800-Cubic-Inch Open Ships			
Gladys O'Donnell	Waco	0:10:45.4	750
Margery Doig	Pitcairn	0:11:05.4	450
Mildred Morgan	Travelair	0:13:59.2	300
Event No. 35: Women's 800-Cubic-Inch Cabin Ships			
Phoebe Omlie	Monocoupe	0:10:43.0	750
Gladys O'Donnell	Monocoupe	0:10:43.6	450
May Haizlip	Cessna	0:11:08.1	300

Pilot	Plane	Time (h:m:s)	Prize ($)

Event No. 36: Women's Dead Stick Landing
28 August

Pilot	Plane	Time (h:m:s)	Prize ($)
Charity Langdon	Whitworth		100
Bettie Lund	Aeronca		50
Nancy Hopkins	Kittyhawk		25

29 August

Mildred Morgan	Travelair		100
Nancy Hopkins	Kittyhawk		50
Bettie Lund	Aeronca		25

30 August

Ruth Nichols	Birdwing		100
Nancy Hopkins	Kittyhawk		50
Bettie Lund	Aeronca		25

31 August

Bettie Lund	Aeronca		100
Ruth Nichols	Brunner Winkle		50
Nancy Hopkins	Kittyhawk		25

1 September

Bettie Lund	Aeronca		100
Mildred Morgan	Travelair		50
Nancy Hopkins	Kittyhawk		25

1931 NATIONAL AIR RACES

Pilot	Plane	Time (h:m:s)	Prize ($)

Transcontinental Handicap Air Derby—Women's Division

Pilot	Plane	Time (h:m:s)	Prize ($)
Phoebe Omlie	Monocoupe		3000
May Haizlip	Monocoupe		1800
Marty Bowman	Inland Sport		1200
Edith Foltz	Bird		
Louise Thaden	Thaden R-4		
Gladys O'Donnell	Waco 10T		
Clema Granger	Swallow		
Mildred Morgan	Travelair		
Joan Fay Shankle	Stearman		
Ruth W. Stewart	Curtiss Robin		

Event No. 24: Women's 350-Cubic-Inch A.T.C. Race

May Haizlip	Davis	0:13:56.40	500
Florence Klingensmith	Monocoupe	0:14:30.60	300
Bettie Lund	Aeronca	0:21:24.62	200

Event No. 25: Women's 510-Cubic-Inch Free-For-All

Phoebe Omlie	Monocoupe	0:13:51.50	500
May Haizlip	Gee Bee D	0:13:54.08	300
Maud Tait	Gee Bee E	0:14:01.56	200

Event No. 26: Women's 650-Cubic-Inch A.T.C. Race

Phoebe Omlie	Monocoupe	0:13:35.21	750
May Haizlip	Gee Bee D	0:13:38.68	450
Maud Tait	Gee Bee E	0:13:43.33	300
Jean LaRene	Rearwin	0:15:00.00	

Pilot	Plane	Time (h:m:s)	Prize ($)

Event No. 27: Women's 800-Cubic-Inch Free-For-All

Pilot	Plane	Time (h:m:s)	Prize ($)
Gladys O'Donnell	Waco	0:11:09.87	1000
May Haizlip	Travelair	0:11:23.77	600
Opal Kunz	Waco	0:11:46.24	400
Florence Klingensmith	Cessna	0:12:03.69	
Bettie Lund	Waco	0:12:27.48	
Margery Doig	Pitcairn	0:12:30.69	
Maud Tait	Gee Bee E	0:13:30.57	
Mildred Morgan	Travelair	0:14:59.41	

Event No. 28: Women's 1000-Cubic-Inch A.T.C. Race

Pilot	Plane	Time (h:m:s)	Prize ($)
Gladys O'Donnell	Waco	0:10:49.17	1250
May Haizlip	Laird	0:10:57.49	750
Florence Klingensmith	Cessna	0:11:14.88	500
Mildred Morgan	Travelair	0:14:21.08	

Event No. 29: Women's 1875-Cubic-Inch A.T.C. Race

Pilot	Plane	Time (h:m:s)	Prize ($)
Florence Klingensmith	Cessna	0:10:37.31	1500
May Haizlip	Laird	0:10:41.99	900
Gladys O'Donnell	Waco	0:11:00.21	600
Joan Shankle	Lockheed	0:11:04.48	
Mildred Morgan	Travelair	Out	

Event No. 33: Women's Free-For-All Aerol Trophy Race

Pilot	Plane	Time (h:m:s)	Prize ($)
Maud Tait	Gee Bee Y	0:15:59.62	3750
May Haizlip	Laird	0:18:09.58	2250
Florence Klingensmith	Cessna	0:18:42.70	1500
Joan Shankle	Lockheed	0:19:35.54	
Phoebe Omlie	Monocoupe	0:20:07.65	
Bettie Lund	Waco TW	0:20:20.85	
Opal Kunz	Travelair	0:20:34.75	
Gladys O'Donnell	Waco	0:20:47.64	

Event No. 35: Women's Dead Stick Landing Contest
31 August (without brakes)

Pilot	Plane	Time (h:m:s)	Prize ($)
Bettie Lund	Aeronca		100
Arlene Davis	Buhl Pup		60
Mary Charles	Swallow		40

1 September (with brakes)

Pilot	Plane	Time (h:m:s)	Prize ($)
Florence Klingensmith	Waco F		100
Edith Foltz	Bird		60
Joan Shankle	Stearman		40

4 September (with brakes)

Pilot	Plane	Time (h:m:s)	Prize ($)
Florence Klingensmith	Waco F		100
Arlene Davis	Buhl Pup		60
Clema Granger	Swallow		40

4–5 September (without brakes)

Pilot	Plane	Time (h:m:s)	Prize ($)
Florence Klingensmith	Waco F		100
Winifred Spooner	Curtiss-Reid		60
Mildred Morgan	Travelair		40

5 September (with brakes)

Pilot	Plane	Time (h:m:s)	Prize ($)
Edith Foltz	Bird		100
Florence Klingensmith	Stearman		60
Bettie Lund	Bird		40

Pilot	Plane	Time (h:m:s)	Prize ($)

6 September (without brakes)

Pilot	Plane	Time (h:m:s)	Prize ($)
Clema Granger	Swallow		100
Arlene Davis	Buhl Pup		60
Florence Klingensmith	Stearman		40

7 September (without brakes)

Pilot	Plane	Time (h:m:s)	Prize ($)
Mildred Morgan	Travelair		160*
Joan Shankle	Stearman		
Edith Foltz	Bird		40

* Morgan and Shankle tied for first place and agreed to divide the $160 prize money.

Event No. 30: Men's and Women's Mixed Race

Pilot	Plane	Time (h:m:s)	Prize ($)
Robert Hall	Gee Bee	0:13:28.54	1250
Jimmy Wedell	Wedell-Williams	0:13:34.30	750
Jim Haizlip	Laird	0:18:16.26	500
Opal Kunz	Travelair	0:20:52.47	
Bettie Lund	Waco TW	Out in 4th lap	
Arthur Davis	Waco 10T	Out in 4th lap	

1932 NATIONAL AIR RACES

Pilot	Plane	Time (h:m:s)	Prize ($)

Cord Cup Race: Transcontinental Handicap Air Derby
Pacific Wing

Pilot	Plane	Time (h:m:s)	Prize ($)
Roy Hunt	Great Lakes		2000
Eldon Cessna	Cessna		500
Art Carnahan	Monocoupe		400
Gladys O'Donnell	Monocoupe		300
J.S. McDonnell	Great Lakes		200
Lloyd O'Donnell	Waco		100
John Hardesty	Monocoupe		100
John B. Vickers	Fairchild		100
Waldo Waterman	Waterman		100
William E. Bleakley	Curtiss Robin		100
Arthur E. Gross	Monocoupe		100
Cecil Allen	Travelair		100
Thomas B. Mullins	Curtiss Robin		100
Kenneth Neese	Travelair		100
Edith Foltz	Bird		100
Ed Bush	Great Lakes		100
Jim Granger	Swallow		100
Marshall Headle	Stearman		100
Leslie C. Miller	Nicholas Beazley		100
Ted Brown	Waco		100
Nixon Galloway	Swallow		100
Neil McGaffey	Bird		100
Henry C. Thompson	Stinson		100
Bob Buck	Pitcairn		100
Fred Burlew	Curtiss Robin		100
Joseph Hager	Curtiss Robin		100
Charles Spencer Jr.	Fleet		
Marty Bowman	American Eagle		
Bob Allen	Waco		
Marion McKenn	Fleet		
Al Lary	Stearman		
Harry Sham	Stearman		
Ross Hadley	Stinson		
Jack Hermann	Travelair		

Pilot	Plane	Time (h:m:s)	Prize ($)
Boyd Grover	Monocoupe		
Frank Reed	Monocoupe		
Ulrich Richter	Klemm		
Clema Granger	Stinson		

Atlantic Wing

Pilot	Plane	Time (h:m:s)	Prize ($)
S.C. Huffman	Waco		1000
Fred Dorsett	De Havilland Moth		500
George C. Lennox	Travelair		400
Helen McCloskey	Monocoupe		300
J.B. Crane	Travelair		200
Harold Neumann	Monocoupe		100
Melville Robinson	Waco		100
Paul Sturtevant	Aristocrat		100
Douglas Davis	Travelair		100
J.A. Field	Travelair		100

William B. Leeds Trophy Race—Roosevelt Field to Cleveland
Men and Women Pilots—Privately Owned Planes Only

Pilot	Plane	Time (h:m:s)	Prize ($)
F.W. Zelcer	Laird		750
Jack H. Wright	Monocoupe		500
Lloyd O. Yost	Waco		250
Bertram J. Goldsmith	Travelair		100
Miss J. Goddard	Monocoupe		50
Leslie B. Cooper	Kellett Autogiro		50
Marcellus A. King	Monocoupe		50
Clyde Pangborn	Pitcairn		50
William Rausch	Gee Bee		50
Carl Dixon	Standard		50
A.S. Fell, Jr.	Travelair		50
Ed A. Voras	Waco		50
Roland Newman	Aero Marine		
Cecil H. Coffrin	Waco		
Blanche Noyes	Travelair		
Victor Pixey	Eaglerock		

Event No. 8: Certificated or Non-Certificated Planes
510-Cubic-Inch Handicap Race

Pilot	Plane	Time (h:m:s)	Prize ($)
R.C. Havens	Taylor Cub	0:13:39.61	180
Florence Klingensmith	Waco	0:13:48.50	100
Jack Morris	Monocoupe	0:13:50.78	60
Roland Newman	Aero Klemm	0:14:05.77	40
Helen McCloskey	Monocoupe	0:14:07.71	20
Paul A. Sturtevant	Aristocrat	0:14:09.71	
Russell Krupp	Monocoupe	0:14:18.56	
E.F. Gallagher	Heath Parasol	0:14:29.03	
Art Davis	Buhl Pup	Out in first lap	

Event No. 10: Certificated or Non-Certificated Planes
125 Miles Per Hour Basis

Pilot	Plane	Time (h:m:s)	Prize ($)
Eldon Cessna	Cessna	0:10:58.52	270
James Herndon	Monocoupe	0:11:34.38	150
Art Chester	Davis	0:11:43.33	90
S.W. Garrigus	Waco	0:12:15.72	60
Joe Meehan	Great Lakes	0:12:30.44	30
Russell Krupp	Monocoupe	0:12:39.93	
Clarence McArthur	Stinson	0:13:36.70	
Annette Gipson	Aristocrat	0:14:40.33	
Roland C. Newman	Stearman	Cut pylon on second lap	

Pilot	Plane	Time (h:m:s)	Prize ($)

Event No. 21: Sohio Mystery Derby—Division No. 1

Pilot	Plane	Time (h:m:s)	Prize ($)
Art Chester	Davis	2:34:00	235
Roy O. Hunt	Great Lakes	2:35:00	141
Lee Sherrick	Travelair	3:01:00	94
J.S. McDonnell	Great Lakes	3:03:00	20
Ethyl Northard	Great Lakes	3:10:00	20
W.P. Jones	Great Lakes	4:01:00	20
William F. Sauters	Bird	3:51:00	20
Helen McCloskey	Pitcairn	3:57:00	20

Event No. 24: Precision Landing Contest—Without Brakes
27 August*

Pilot	Plane	Prize ($)
E.F. Beckley	Taylor Club	45
Winston Kratz	Aeronca	25
Harold Neumann	Monocoupe	15
Art Davis	Buhl Pup	10
A.A. Walzak	Bird	5
John Barstow	Stinson	
Franklin Knapp	Avro Avian	
Helen McCloskey	Monocoupe	
Walter J. Carr	Carr Special	

* Competition was scheduled for 27 August but was held on 28 August. Women pilots also entered women-only events.

Event No. 22: Aerol Trophy Race—Free-For-All

Pilot	Plane	Time	Prize ($)
Gladys O'Donnell	Howard	0:12:56.38	2250
May Haizlip	Wedell-Williams	0:13:06.34	1250
Florence Klingensmith	Monocoupe	0:13:47.78	750
Bettie Lund	Waco	0:23:39.20	500

Event No. 23: Amelia Earhart Trophy Race—685-Cubic-Inch Handicap

Pilot	Plane	Time
Florence Klingensmith	Monocoupe	0:15:07.39
Edith Foltz	Bird	0:15:31.99
Helen Richey	Bird	0:15:36.78
Mary M. Sansome	Fleet	0:15:47.88
Helen McCloskey	Monocoupe	0:16:21.47
Rae Trader	Trader Special	0:17:14.01

Event No. 29: Shell Petroleum Corp. Speed Dashes—Women Only for World's Record Over Three-Kilometer Course

Pilot	Plane	Time	Prize ($)
May Haizlip	Wedell-Williams	255.513 (mph)	675
Florence Klingensmith	Monocoupe	198.611	375
Bettie Lund	Waco	179.795	225

1933 NATIONAL AIR RACES

Pilot	Plane	Time (h:m:s)	Prize ($)

Event No. 9: Aerol Trophy Race—Free-For All

Pilot	Plane	Time	Prize ($)
May Haizlip	Wedell-Williams	0:17:50.04	1350
Marty Bowman	Gee Bee	0:18:33.12	750
Gladys O'Donnell	Waco	0:22:23.24	450
Henrietta Sumner	Travelair Speedwing	0:23:05.53	300

1935 NATIONAL AIR RACES

Pilot	Plane	Time (h:m:s)	Prize ($)
Ruth Chatterton Sportsman Pilot Trophy Race			
Grace E. Prescott	Travelair		450
W.S. Woodson	Fleet		250
Leland Hayward	Waco		150
Cecile Hamilton	Aeronca		100
Ethel A. Sheehy	Great Lakes		50
Event No. 4: Women's A.T.C. Race—Free-For-All			
Edith Berenson	Bird	0:18:18.78	562.50
Melba Beard	Bird	0:18:31.60	312.50
Edna Gardner	Porterfield	0:18:38.94	187.50
Genevieve Savage	Great Lakes	0:18:51.58	125.00
Peggy Remey	Travelair	0:18:52.48	62.50

1936 NATIONAL AIR RACES

Pilot	Plane	Time (h:m:s)	Prize ($)
Amelia Earhart Trophy Race—A.T.C. Race			
Betty Browning	Cessna	0:15:58.62	675
Gladys O'Donnell	Ryan	0:16:10.01	375
Genevieve Savage	Ryan	0:16:27.25	225
Jeannette Lempke	Davis	0:16:30.94	150
Nancy Love	Beechcraft	0:16:44.06	75
Henrietta Sumner	Cessna	0:16:44.58	
Edna Gardner	Porterfield	0:16:47.29	
Melba Beard	Porterfield	0:16:58.08	
Ruth Chatterton Sportsman Pilot Trophy Race *Eastern Wing—Cleveland to Dallas*			
Frank Spreckels	Luscombe		250
George Arents	Stinson		150
Jeannette Lempke	Davis		100
Clara E. Livingston	Stinson		100
Helen McCloskey	Monocoupe		50
Western Wing—Dallas to Los Angeles			
George S. Armistead	Stinson		300
Max Marshall	Fairchild		180
Jeannette Lempke	Davis		120
Jerry Fairbanks	Stinson		120
Grace Prescott	Stinson		120
Sweepstakes Winners			
Frank Spreckels	Luscombe		Chatterton Trophy
Jeannette Lempke	Davis		Stinson Award
Clara Livingston	Stinson		Seventh place

Also participating in the 1936 Chatterton Derby were Grace Anderson, Adrienne Clark, Edith Clark, Katherine Sui Cheung, Majorty Jane Gage, Edna Gardner, Annette Gipson, Cecile Hamilton, Evelyn Hudson, and Peggy Salaman.

1937 NATIONAL AIR RACES

Pilot	Plane	Time (h:m:s)	Prize ($)
Amelia Earhart Memorial Trophy Race			
Gladys O'Donnell	Ryan	0:11:34.16	450
Betty Browning	Cessna	0:12:22.57	250
Edna Gardner	Cessna	0:12:59.05	150
Annette Gipson	Monocoupe	0:13:19.56	100
Dorothy Munro	Rearwin	0:16:52.45	50
C.G. Taylor Light Plane Derby			
E.H. Spiller	Taylor-Young		150
W. Graham	Taylor		100
H. Carl McQuigg	Aeronca K		75
R.H. Bell	Taylor Cub		60
Jack Marchand	Aeronca		55
Abbie Dill	Taylor		50
J.B. McLester	Taylor		45
L.R. Furlong	Taylor		40
D.D. Frye	Taylor Cub		25
W. Kenyon	Taylor Cub		10
C.A. Dixon	Aeronca		10
Ione Coppedge	Aeronca		10
John Jones	Taylor		10
R.L. Stout	Taylor-Young		10
Arlo Mather	Taylor Cub		10

INTERNATIONAL AIR RACES, CHICAGO, 1933

Pilot	Plane	Speed (mph)	Prize ($)
Women's International Free-For-All			
May Haizlip	Wedell-Williams	191.11	1125
Florence Klingensmith	Gee Bee	189.04	625
Marty Bowman	Gee Bee	168.86	375
Henrietta Lantz	Howard Special	123.11	250

FIRST WOMEN'S NATIONAL AIR MEET 1934

Pilot	Plane	Time (h:m:s)	Prize ($)
20-Mile Free-For-All Handicap Race, 4 August			
Jeanette Lempke	Great Lakes		100*
Annette Gipson	Aristocrat		50
Gladys O'Donnell	Monocoupe		25

* Trophy for first place: silver bowl, donated by Dayton Rubber Manufacturing Co.

Precision Landing Contest, 4 August			
Ellen Smith	Waco		40*
Mrs. C.F. Kolp	Spartan		30
Ione Coppedge	Aeronca		15
Annette Gipson	Aristocrat		10
Edna Gardner	Waco Taperwing		5

* Trophy for first place: silver tray, donated by Pratt & Whitney.

Bomb Dropping Contest, 5 August			
Mrs. C.F. Kolp	Spartan		40*
Mabel Britton	Waco F		30
Edna Gardner	Waco Taperwing		15
Ione Coppedge	Aeronca		10
Jeannette Lempke	Great Lakes		5

* Trophy for first place: cigarette case and lighter, donated by *Dayton Daily News*.

50-Mile Free-For-All Handicap Race, 5 August			
Helen Richey	Eaglerock	0:38.53	1000
Arlene Davis	Stinson R	0:39.00	500
Gladys O'Donnell	Monocoupe	0:39.10	250
Helen McCloskey	Monocoupe	0:39.20	150
Ruth Nason	Waco F	0:39.36	100

* Trophy for first place donated by Fairchild Aircraft Corp., and a Sperry artificial horizon.

ALL AMERICAN AIR MANEUVERS, MIAMI
10–12 January 1935

Pilot	Plane	Speed (mph)
Women's Handicap Race		
Edna Gardner Kidd	Aristocrat	98.16
Women's Aerobatics		
Helen McCloskey	Monocoupe	

ALL AMERICAN AIR MANEUVERS, MIAMI
6–8 January 1939

Pilot
K.K. Culver Trophy—Free-For-All Handicap Race—50 Miles
Edna Gardner Kidd
Edith Descomb
Bessie Owens
Florence H. Boswell

Appendix 2

Records Set by U.S. Women in the 1930s

International Records

HEAVY LANDPLANES

Distance in Straight Line without Refueling and without Load

Ruth Nichols, Lockheed Vega, Pratt & Whitney Wasp 640 hp engine, 24–25 October 1931. Distance: 1977.6 miles.
Amelia Earhart, Lockheed Vega, Pratt & Whitney Wasp 450hp engine, 24–25 August 1932. Distance: 1477.728 miles.

Duration with Refueling in Flight

Evelyn (Bobbi) Trout and Edna May Cooper, Curtiss Robin, 4–9 January 1931. Duration: 123 hours.
Louise Thaden and Frances Marsalis, Curtiss Thrush, Wright J6-E engine, Vallery Stream, New York, 14–22 August 1932. Duration: 196 hours, 5 minutes.
Frances Marsalis and Helen Richey, Miami, Florida, 20–30 December 1933. Duration: 237 hours, 42 minutes.

Altitude without Payload

Elinor Smith, Bellanca monoplane, Wright J5 200hp engine, 10 March 1930. Height: 27,418.317 feet.
Ruth Nichols, Lockheed Vega, Pratt & Whitney Wasp engine, 6 March 1931. Height: 28,743.352 feet.

Maximum Speed over a 3-Kilometer Straightaway Course without Payload

Amelia Earhart, Lockheed Vega, Pratt & Whitney Wasp engine, 5 July 1930. Average speed: 181.157 miles per hour.
Ruth Nichols, Lockheed Vega, Pratt & Whitney Wasp engine, 13 April 1931. Average speed: 210.704 miles per hour.
May Haizlip, Wedell-Williams, Pratt & Whitney Wasp Junior 540hp engine, Cleveland, Ohio, 5 September 1932. Speed: 252.226 miles per hour.
Jacqueline Cochran, modified Seversky p-35, Pratt & Whitney 850hp engine, Detroit, Michigan, 21 September 1937. Speed: 292.271 miles per hour.

Speed for 100 Kilometers in Closed Circuit without Payload

Amelia Earhart, Lockheed Vega, Pratt & Whitney Wasp 420hp engine, Detroit, Michigan, 25 June 1930. Speed: 174.897 miles per hour.

Speed for 100 Kilometers in Closed Circuit with Payload of 500 Kilograms

Amelia Earhart, Lockheed Vega, Pratt & Whitney Wasp 420hp engine, Detroit, Michigan, 25 June 1930. Speed: 171.438 miles per hour.

LIGHT LANDPLANES

Multi-Seaters Having an Empty Weight of Less than 1235 Pounds
Distance without Load

Helen McCloskey and Mrs. Monroe McCloskey, Monocoupe, Lambert 90hp engine, 25 June 1936. Distance: 524.126 miles.

Speed for 100 Kilometers

Helen McCloskey and Genevieve Savage, Monocoupe, Warner 145hp engine, Miami, Florida, 15 January 1935. Speed: 166.632 miles per hour.

Single-Seaters Having an Empty Weight of Less than 992 Pounds
Speed for 100 Kilometers

Louise Thaden, Porterfield monoplane, Warner 90hp engine, 12 July 1936. Average speed: 109.58 miles per hour.
Annette Gipson, Monocoupe, Lambert 90hp engine, 30 July 1936. Average speed: 123.247 miles per hour.

Single-Seaters Having an Empty Weight between 440 and 771 Pounds
Altitude

May Haizlip, Buhl Pup, Szekeley 85hp engine, St. Clair, Michigan, 13 June 1931. Height: 18,097 feet.

Multi-Seaters Having an Empty Weight of Less than 617 Pounds
Altitude

Ione Coppedge and Josephine Garrigus, Aeronca, Aeronca 36hp engine, Dayton, Ohio, 11 February 1936. Height: 15,253 feet.

Speed for 100 Kilometers

Helen Frigo, Aeronca C-3, Aeronca 36hp engine, College Park, Maryland, 12 June 1936. Speed: 74.193 miles per hour.

Single-Seaters Having an Empty Weight of Less than 441 Pounds
Altitude

Helen Richey, Aeronca C-2, Aeronca 36hp engine, 9 May 1936. Height: 18,448 feet.
Irene Crum, Aeronca C-2, Aeronca E-113-C 36hp engine, 1 February 1937. Height: 19,426 feet.

Speed for 100 Kilometers

Helen Richey, Aeronca C-2, Aeronca 36hp engine, Hampton, Virginia, 1 February 1936. Speed: 72.224 miles per hour.

HEAVY SEAPLANES

Altitude

Mary Eddy Conrad, Savoia-Marchetti, Kinner 125hp engine, Port Washington, New York, 20 October 1930. Height: 13,461 feet.

LIGHT SEAPLANES

Multi-Seaters Having an Empty Weight of Less Than 1499 Pounds
Altitude

Crystal Mowry and Alice Bender, Kittyhawk, Kinner 125hp engine, 12 December 1936. Height: 6070 feet.

Speed for 100 Kilometers

Crystal Mowry and Edith McCann, Kittyhawk, Kinner 125hp engine, 9 December 1936. Speed: 79.138 miles per hour.

Single-Seaters Having an Empty Weight of Less than 1257 Pounds
Speed for 100 Kilometers

Margaret Bain Tanner, Aeronca seaplane, Aeronca 36hp engine, 8 August 1936. Speed: 66.672 miles per hour.

Multi-Seaters Having an Empty Weight of Less than 772 Pounds
Altitude

Crystal Mowry and Lillian Bishop, Aeronca seaplane, Aeronca 36hp engine, Miami, Florida, 10 December 1936. Height: 5830 feet.

Speed for 100 Kilometers

Crystal Mowry and Edith McCann, Aeronca seaplane, Aeronca 36hp engine, Miami, Florida, 10 December 1936. Speed: 69.011 miles per hour.

Women's National Records

HEAVY LANDPLANES

Altitude without Load

Jacqueline Cochran, Beechcraft Staggerwing, Pratt & Whitney 600 hp engine, Palm Springs, California, 24 March 1939. Height: 30,052.430 feet.

Speed for 100 Kilometers

Louise Thaden, Beechcraft Staggerwing, Wright Whirlwind 450 hp engine, St. Louis, Missouri, 29 May 1937. Speed: 197.958 miles per hour.

Jacqueline Cochran, Beechcraft Staggerwing, Pratt & Whitney Wasp 600 hp engine, 28 July 1937. Speed: 200.712 miles per hour.

Jacqueline Cochran, Seversky P-35, Pratt & Whitney Twin Row Wasp, Miami, Florida, 13 December 1937. Speed: 255.942 miles per hour.

Speed for 1000 Kilometers

Jacqueline Cochran, Beechcraft Staggerwing, Pratt & Whitney Wasp 600hp engine, 26 July 1937. Speed: 203.895 miles per hour.

LIGHT LANDPLANES

Altitude

Grace Huntington, Fairchild 24, Warner Super Scarab engine, 31 May 1939. Height: 18,769.646 feet.

Annette Gipson and Mrs. John Buckman, Monocoupe, Lambert 90 hp engine, Ft. Lauderdale, Florida, 26 April 1936. Height: 12,628 feet.

AUTOGIROS

Altitude

Amelia Earhart, Pitcairn, Wright J6-9 300 hp engine, Willow Grove, Pennsylvania, 7 April 1931. Height: 18,415 feet.

DIESEL-POWERED AIRCRAFT

Altitude

Ruth Nichols, Lockheed Vega, Packard Diesel 225 hp engine, Floyd Bennett Field, New York, 14 February 1932. Height: 19,928 feet.

GLIDERS

Duration (single-place)

Helen M. Montgomery, Stevens-Franklin, Crystal Downs Beach, Frankfort, Michigan, 4 September 1938. Duration: 7 hours, 28 minutes.

Women's International Continental and Inter-City Distance Records

East-to-West Flights

Ruth Nichols, Lockheed Vega, Mineola, New York, to Burbank, California, 24 November–1 December 1930. Total flying time: 16 hours, 59 minutes, 30 seconds.

Laura Ingalls, Lockheed Orion, Pratt & Whitney Wasp 500hp engine, Floyd Bennett Field, Brooklyn, New York, to Union Air Terminal Burbank, California, 11 July 1935. Total elapsed time: 18 hours, 23 minutes.

Louise Thaden and Blanche Noyes, Beechcraft Staggerwing, Wright 420hp engine, Floyd Bennett Field, Brooklyn, New York, to Los Angeles, California, 4 September 1936. Total elapsed time: 14 hours, 55 minutes, 1 second.

West-to-East Flights

Ruth Nichols, Lockheed Vega, Burbank, California, to Mineola, New York, 9–10 December 1930. Total flying time: 13 hours, 21 minutes, 43 seconds.

Laura Ingalls, Lockheed Orion, Pratt & Whitney Wasp 500hp engine, Burbank, California, to Floyd Bennett Field, Brooklyn, New York, 12 September 1935. Total elapsed time: 13 hours, 34 minutes, 5 seconds.

Jacqueline Cochran, Seversky P-35, Pratt & Whitney Twin Row Wasp engine, Union Air Terminal, Burbank, California, to Floyd Bennett Field, Brooklyn, New York, 3 September 1938. Total elapsed time: 10 hours, 27 minutes, 55 seconds.

Los Angeles to Mexico City

Amelia Earhart, Lockheed Vega, Pratt & Whitney Wasp 500hp engine, 19–20 April 1935. Total elapsed time: 13 hours, 33 minutes.

Mexico City to Newark

Amelia Earhart, Lockheed Vega, Pratt & Whitney Wasp 500hp engine, 8 May 1935. Total elapsed time: 14 hours, 19 minutes.

New York City to Miami

Jacqueline Cochran, Seversky P-35, Pratt & Whitney Twin Row Wasp engine, 3 December 1937. Total elapsed time: 4 hours, 12 minutes, 27.2 seconds.

Detroit to Akron

Louise Thaden, Beechcraft Staggerwing, Wright 420hp engine, 21 January 1937. Total elapsed time: 40 minutes, 43 seconds.

References

Ackermann-Blount, Joan
1984. She Flies Through the Air With the Greatest of Ease. *Sports Illustrated* (16 January), 60(2):102–104.
Albert, Dora
1930. Bessie Davis, Aviation's Flying Saleswoman. *The American Magazine* (June), 109:88.
Author unknown
1930a. Women Take to the Air. *U.S. Air Services* (May), 15(5):44.
1930b. Three Wings—America's Best. *Newark Free Press* (24 September).
1930c. Women's Work. *Western Flying* (December), 7(12):33.
1930d. "New York Girl Makes 714 Barrel Rolls," unidentified clipping (14 August) in Laura Ingalls biographical file, National Air and Space Museum archives, Smithsonian Institution, Washington, D.C.
1931a. 2 Girl Endurance Flyers Soar on in Cloudy Skies. *Newark Star-Eagle* (5 January).
1931b. Girl Flyers Land after 122 Hours Aloft. *New York Times* (10 January).
1931c. Girl Aviatrix Faints in Air, Hurts Slight. *Newark Star-Eagle* (27 March).
1931d. The Flier's Own Story. *New York Times* (10 April).
1933. Treasure Hunting on Wings. *Liberty* (23 September), 10(38):25.
1934. This Month's Cover. *U.S. Air Services* (June), 19(6):32.
1937. Woman Pilots on Police Duty. *Los Angeles Examiner* (10 May).
1938. Judging from the Records There's No Such Thing as "Only Woman Engineer." *American Aviation* (15 August), 2(6):13.
1939. Woman Passenger Sets Mark in 16 Day Flight Around World. *New York Times* (16 July).
1975. Florence Barnes, 20's Aviator, Dies. *New York Times* (31 March).
Backus, Jean L.
1982. *Letters From Amelia*. Boston: Beacon Press.
Brown, Margery
1930. Flying is Changing Women. *Pictorial Review* (June), 31:30.
Calkins, Selby
1939. Amazing Miss. *Popular Aviation* (July),25(1):49–50.
Cochran, Jacqueline
1954. *The Stars at Noon*. Boston: Little, Brown, and Co.
1979. Interview with Claudia M. Oakes (20 May).
Cowen, Ron
1983. Fearless Flying ... and Falling. *Washington Post* (5 November), C2.
Crouch, Tom D.
1983. *The Eagle Aloft*. Washington, D.C.: Smithsonian Institution Press.
Driscoll, Helen
1931. Betty Lund Banks on "13" Aid in Races. *Cleveland Press* (29 August).
Dwiggins, Don
1965. *They Flew the Bendix Race*. Philadelphia and New York: J.P. Lippincott Co.
Earhart, Amelia
1930a. Women's Influence on Air Transport Luxury. *The Aeronautic Review* (March), 8(3):32.

1930b. Women's Influence on Aviation. *The Sportsman Pilot* (April), 3(4):15.
Gillies, Betty H.
1979. Letter to Claudia M. Oakes, 16 May, in files of C.M. Oakes, Department of Aeronautics, National Air and Space Museum, Smithsonian Institution, Washington, D.C.
Griffin, Clementina de Forest
1938. Women Technicians—Why Not? *N.A.A. Magazine* (15 August), 2(6):17.
Hager, Alice Rogers
1939. I Flew For a Week. *Popular Aviation* (January), 25(1):27–28.
Hyman, Helen
1932. Bettie Lund Seeks To Prove Flying is Safe. Unidentified newspaper clipping (1 April) in Bettie Lund biographical file, National Air and Space Museum archives, Smithsonian Institution, Washington, D.C.
Johnson, Helen R.
1935. We Also Instruct. *Southwestern Aviation* (June), 8(6):10.
Johnson, Jesse J., editor
1974. *Black Women in the Armed Forces 1942–1974*. Hampton, Va.: Hampton Institute.
Kerfoot, Glenn
1978. Helen Richey, First Lady of the Airlines. *Ninety-Nine News* (April), 5(4):17–18.
Lindbergh, Anne Morrow
1935. *North to the Orient*. New York: Harcourt, Brace, and World, Inc.
Mandrake, Charles G.
1957. *The Gee Bee Story*. Wichita: R.R. Longo Co.
Maravelas, Paul
1980. Jeannette Piccard Interviewed. *Ballooning*, 13(4):17.
Moolman, Valerie
1981. *Women Aloft*. Alexandria, Va.: Time-Life Books.
Nichols, Ruth
1932. Behind the Ballyhoo. *The American Magazine* (March), 113:80–83.
Nye, Harriet D.
1982. Letter to Claudia M. Oakes, 1 November, in files of C.M. Oakes, Department of Aeronautics, National Air and Space Museum, Smithsonian Institution, Washington, D.C.
Roberts, Mary M.
1931. Letter to Robert Johnson, 8 July, in files of C.M. Oakes, Department of Aeronautics, National Air and Space Museum, Smithsonian Institution, Washington, D.C.
Schimmoler, Loretta M.
1939. Aerial Amazons? *National Aeronautics* (August), 17(8):34.
Studer, Clara
1935. Bread and Butter and Airwomen. *National Aeronautics* (October), 13(9):20.
Sutton, Leslie
1970. Louise Thaden Recalls Half-Century of Petticoat Flying. *Northwest Arkansas Times* (25 August), page 3,column 3.
Thaden,Louise M.
1931. Training Women Pilots. *Western Flying* (Febuary), 9(1):22.

1936. Five Women Tackle the Nation. *N.A.A. Magazine* (August), 14(8):14–15.

1938. *High, Wide, and Frightened.* New York: Stackpole Sons.

1979a. Letter to Claudia M. Oakes, 25 April, in files of C.M. Oakes, Department of Aeronautics, National Air and Space Museum, Smithsonian Institution, Washington, D.C.

1979b. Telephone interview with Claudia M. Oakes (4 May).

Thompson, Virginia

1970. Backward Glance Column. *Ninety-Nine News* (March), 14(16):6.

Trout, Bobbi

1983. Letter to Claudia M. Oakes, 5 July, in files of C.M. Oakes, Department of Aeronautics, National Air and Space Museum, Smithsonian Institution, Washington, D.C.

1930. The Mind of an Air Girl. *Psychology* (July), 15(1):18.

Wiggins, J.E.

1930. Women and Pilots and Women Pilots. *Pacific Flyer* (December), 5(6):9.

Williams, Edwin M.

1932. How Phoebe Omlie Won the 1931 Air Derby. *Southern Aviation* (April), 3(8):15–16, 31.

Wittman, Charles

1975. It Started as an Experiment. *Times-Plain Dealer,* Casco, Iowa (9 July), 109(28).